Digital Electronics:

A Hands-On Learning Approach

GEORGE YOUNG

HAYDEN BOOK COMPANY, INC.
Rochelle Park, New Jersey

Library of Congress Cataloging in Publication Data

Young, George, 1928–
  Digital electronics, a hands-on learning approach.

  SUMMARY: A textbook of digital electronics
featuring almost exclusively an experimental or
laboratory approach.
  Includes index.
  1. Digital electronics.   [1. Digital electronics]
I. Title.
TK7868.D5Y68   621.381   80–17388
ISBN 0–8104–5668–0

Printed in the United States of America

        4   5   6   7   8   9   PRINTING

    82 83 84 85 86 87 88   YEAR

Preface

The traditional approach to learning electronics consists of lectures on theory, some lab work, and written tests. As a rule, a great deal of time is spent expounding theory and testing the student's grasp of it, and considerably less time is spent on lab work. An education of this kind supplies industry with knowledgeable employees who can do very little, the exact opposite of the kind of employees that an employer seeks. Granted that the more an employee knows, the more he should be able to produce. Unfortunately, this is not always the case.

There are many ways to learn electronics. One method rejects the more conventional emphasis on theory and student feedback in favor of an almost exclusively experimental or laboratory approach. Theory is stressed only when the student needs it to make a circuit function. This is the approach used in this text. It will not make electronics any easier to learn, nor will it produce more knowledgeable graduates. It will produce a self-satisfied employee who can produce.

Those of us who have been in the electronics field for any length of time have lost the ability to appreciate the newcomer's bewilderment. We forget that we don't speak "English" in electronics. For this reason, you may find the introductory material almost *too* elementary. Patience, things will get rough very quickly!

The text is accumulative. What is learned in the initial experiments provides the basis for what comes later. As in the study of algebra, failure to master the initial material will prove disastrous within a very short time.

The first experiment (Fig. 1-1), called "Putting Fidgety Fingers to Work," is a buffer for the instructor. Its purpose is to "buy" the instructor a little time at the start of the semester since he has hundreds of things to do during the first few days of classes. He can assign the first couple of chapters (there is a Chapter 0 as well as a Chapter 1) and allow the students to get their feet wet with this experiment while he handles the paperwork chores. The experiment is repeated in Chapter 3 after some additional basic background has been developed. The student using this text at home should read Chapters 0 and 1, skip the first experiment, and go on to Chapter 2.

The instructor can also use the first experiment to ascertain the electronics background of his transfer students. Some students will sail through the first experiment without batting an eye, whereas others will flounder helplessly until their background has been improved. In the classroom, do not hesitate to let those students coming to you with some background help those with no background.

There are no questions or problems in this textbook for the student to work on. To forestall any criticism on this point, I want to inform you that their omission was deliberate.

Most school systems make students take written tests for all their school years. They do not need another class requiring written tests. (They will not be taking tests on their jobs.) I am sure that your administration will not let you get away with not evaluating student progress. Give them a *performance* test instead of a written test. Have them show you their operational circuit. Ask them a few questions about how and why it works. Ask them what would happen if this resistor were doubled in size or that capacitor were replaced with one with half the value. Then have them make these changes.

The study of microprocessors and computers in a beginning text may seem out of place, but it has a purpose. Show the students the microprocessor or computer chapters and their circuit diagrams. Tell them that before they get to these chapters they will be able to read and understand the circuit diagrams. They won't believe you, of course, but when the time comes, they will surprise themselves with their ability to do so.

Too many texts leave out the "hard" circuits and relegate them to "the next level" of a student's education. Is it not preferable to have the "big picture" available for them from the beginning so that they can see where all their effort is leading?

This text makes use of a fairly recent development in electronics—the solderless breadboard. The solderless breadboard recommended by the author—called a "Superstrip"—is manufactured by AP Products, Inc. in two versions. The one with nickel-plated contacts is less expensive than the "top of the line" model, which has gold-plated contacts. Either model will prove satisfactory for the purposes of this text. Other versions of the solderless breadboard may be substituted with more or less satisfactory results.

Digital electronics is easier to learn than analog electronics. Once the fundamentals of the former have been mastered, the step-up to analog electronics becomes much easier. Some analog devices will be studied in this text because certain linear devices are necessary to make the digital devices operate.

Although electronics is one of the more difficult areas of learning to master, it can also be a lot of fun. Good luck!

GEORGE YOUNG

Contents

Chapter ∅

How To Use This Book

What's this? A chapter numbered "∅"? There's a reason. In the field of digital electronics and in the world of microcomputers, the counting process most often starts with the numeral zero instead of the numeral one. Picking up the cue, we have a chapter zero to start with instead of a chapter one.

The Solderless Breadboard

This book is an experimental approach to the learning of electronics. The heart of the book is based on a relatively recent innovation in the field called the *solderless breadboard.*

The solderless breadboard we recommend is one made by AP Products, Inc.[1] This device, called a "Superstrip," is available in two versions, a gold-plated model and a nickel-plated model. The gold-plated version is naturally more expensive. Either version will be satisfactory for the purposes of this text.

These solderless breadboards are possibly the "Cadillac" of the breadboard line. Any of the many solderless breadboards available may be substituted, but the experiments in this text were set up for the "Superstrips," and some juggling of the circuits may be necessary with other solderless breadboards.

Figure ∅-1 shows the AP Products Superstrips in a front view; Fig. ∅-2 shows a rear view of the buss structure inside the plastic housing. In the latter, the self-adhesive backing material has been removed to allow you to see the buss structure. Please note that the contacts down the central portion of the Superstrip run at right angles to its length. Each clip connects five plug-in holes together on the front surface. Connections are made by

---

[1] The address of AP Products, Inc., is Box 110, Painesville, Ohio 44077. The toll-free telephone number is 800-321-9668. Superstrips are also available throughout the U.S. from franchised AP Products distributors. You may determine the nearest distributor with a phone call to the factory in Painesville.

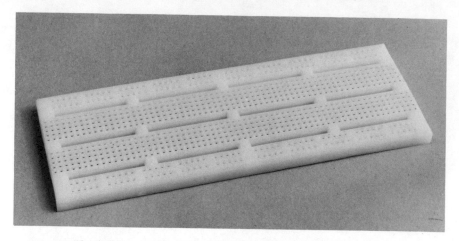

Fig. 0-1 Front view of Superstrip (*Courtesy,* AP Products, Inc.)

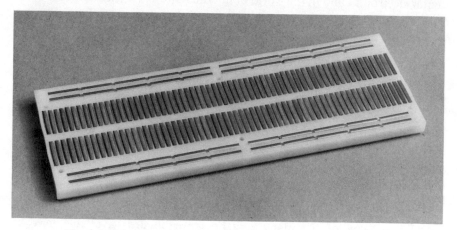

Fig. 0-2 Rear view of Superstrip (*Courtesy,* AP Products, Inc.)

inserting electronic hook-up wire into any one of the five holes, and all five holes connect the same point together.

The busses running parallel to the length of the Superstrip are the power distribution clips, or *power rails*. These longer contacts are used to distribute the power and ground connections to the circuitry. Note in particular that the rails are split into two sections near the center of the Superstrip. In order to have continuous power available along the entire length of the rail on each side, the gap in the center must be bridged with a jumper wire.

For the majority of our experiments, we will assign the two innermost rails on each side to ground or power supply minus, and the two outer rails along each side to + 5 V.

Additional Materials

Many additional materials are needed. Appendix C lists all those needed for each experiment. Only some general requirements will be discussed here.

In addition to the solderless breadboard, we will need a source of power for our experiments. If you are working in a school lab, this power may be obtained from the lab power supplies. However, if you are working with this text at home, then we are going to get you started using a 6-V lantern battery.

A number of jumper wires made from #22 or #24 solid hook-up wire are also needed. All this wire may have the same color insulation, but a variety of colors will prove easier to use and will speed the educational process. Telephone cable scrounged from your friendly telephone installer can be an inexpensive and suitable source for this insulated, solid hook-up wire. Since the wire can be used over and over again, it should be saved after the completion of each experiment.

In addition to these basic components, resistors, capacitors, potentiometers, diodes, LEDs, transistors, integrated circuits, and the like will be necessary. The listing of such components in the Appendix is meant to accommodate the pocketbooks of as many readers as possible. Educational institutions may use this appendix for budgeting purposes. Readers with fatter wallets may use it to make purchases in larger quantities. Those of more modest means may salvage or scrounge components from electronic equipment that has outlived its usefulness, thereby achieving considerable cost savings. Components may be salvaged from transistor radios, old TVs, old cassette players and recorders, inoperational stereos, and so on. Every little bit saved will add up to a sizeable savings in the overall picture. Empty milk cartons, after being washed out and dried, can be cut off at a uniform height to serve as very inexpensive storage devices.

General Plan Of Experiments

Each experiment in this book takes an electronic component and utilizes a circuit to test it. Then the component is utilized in a circuit to make it do something. Gradually more and more components are interconnected in more and more complex circuits as your knowledge of electronics expands.

Try to make each of the experiments work before you proceed to the next. A chapter on troubleshooting is included at the end. Use this chapter as often as you need it. If you make no errors, you aren't human. Human beings make lots of errors, and I have learned from many years of teaching that the more errors you make and correct, the more you will learn!

You will note that there are no questions to answer at the end of each chapter as you would expect to find in a textbook. There's a good reason for that. What we are after here is *performance* on the part of the participant.

In the classroom, the instructor will have to check the performance of the student. This will cause a bit of extra work for the instructor, but students really appreciate not having written tests accompanying each chapter. They demonstrate their newly earned knowledge by performance.

The reader at home would most likely have glanced at the questions (if they had been included) and gone on to the next experiments anyway. Consequently, I do not feel as though I am shortchanging this reader either. The fact that he can make the experiments work, learns how to read schematic diagrams, and masters the symbols of the different electronic components is satisfaction enough. The reader can ask himself "What would happen if . . . (I changed this component to a different value, reversed this or that connection, or . . ."). The reader can then proceed to get the answer to his question by trying out the circuit change on the breadboard. The more the individual puts into experimental work, the greater his rewards.

Electronics can be a lot of fun. It can also be extremely frustrating. When your efforts cease to be fun, stop. Do something else for a while and then return.

Chapter 1

The Power Sources

Every piece of electronic equipment requires a source of power for its operation. This power is usually direct current (dc), which may be obtained from batteries or from a power supply. A power supply is a device that operates from the alternating-current (ac) mains and by means of suitable electronic components and circuitry furnishes the required dc power to operate the equipment.

To get us started, we can use batteries. Batteries produce electrical power by a chemical reaction. If the chemical reaction is not reversible (the materials that make up the battery are "used up" in producing the electrical power), the battery is called a *primary battery* or *primary cell.* On the other hand, some batteries can be recharged and used over and over again. (The chemical reaction *is* reversible and the materials that make up the battery are not "used up" but only changed.) Such batteries are called *secondary batteries* or *secondary cells.* An appropriate charger is needed for the recharging.

Flashlight cells are primary batteries. The chemical reaction is not reversible. As the battery produces electrical power, the materials inside the battery are "used up" in the chemical reaction. When the materials are all "used up," the cell is exhausted and must be discarded and replaced with a new battery.

Most batteries are made up from a number of cells. Each cell produces a specific voltage. Groups of individual cells may be connected together in *series* to produce higher voltages, and the end voltage will be the sum of the number of individual cells used to make up the battery. An ordinary *dry cell* or flashlight battery has a voltage of about 1.5 V. Two of these cells connected in series will produce 3.0 V. Four of these cells connected in series will produce 6.0 V. The ordinary lantern battery has four flashlight cells connected in series to produce an output voltage across its plus and minus terminals of 6.0 V.

5

A nickel cadmium battery is a secondary battery; that is, it may be recharged. Each nickel cadmium cell has a voltage of 1.25 V. Four of these cells connected in series will produce 5.25 V across the plus and minus terminals. To use a NiCd battery, you must also have a suitable charger.

The simplest way for us to get started with our experiments is to use a 6-V lantern battery. Four ordinary dry cells such as "D". flashlight batteries may also be connected in series to provide a suitable power source.

Most of our experiments will deal with Transistor-Transistor-Logic Integrated Circuits (TTL ICs). These devices use +5 V for their operation. The range of voltages over which they will operate correctly is 4.75 to 5.25 V. Thus, we must find a means of lowering the 6 V of our lantern battery to this range. We can use a series-connected silicon diode for this purpose.

Putting Fidgety Fingers to Work

The circuit for this experiment is given in Fig. 1-1(a); Fig. 1-1(b) illustrates the physical set up of the solderless breadboard. This basic circuit will be repeated again in Chap. 3 and will be explained in more detail there. Connect power and ground *first*. The +5 V goes to pin 14, and the ground

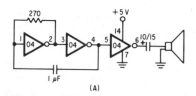

(A)

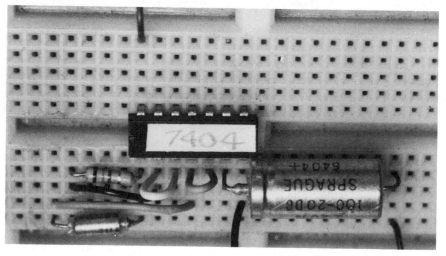

(B)

Fig. 1-1 (a) Circuit diagram for putting "fidgety fingers" to work, and (b) physical layout of the solderless breadboard.

on the power source goes to pin 7. The notch on the integrated circuit (IC) is between pins 1 and 14. Some ICs have a dot or dimple adjacent to pin 1 on the upper surface of the IC. The pin numbers are read counterclockwise (looking at the top of the IC). A straight line on the schematic represents a wire. The wire should be insulated, solid hook-up wire about number 22 or 24 gauge. The ends should be stripped by about ¼ in. for insertion into the holes on the solderless breadboard. The resistor (device with the colored bands on it) may be connected into any unused breadboard locations and then connected to the appropriate pins on the IC with short jumper wires. It is not necessary to bend the resistor leads into a pretzel to make the connections between pins 1 and 2. The capacitors likewise may be located anywhere on the solderless breadboard and connected into the circuit with jumper wires.

When operational, the circuit will emit a fairly loud tone in the speaker. In classrooms, the first student to get his "squawker" going will trigger a chain reaction in the classroom, and other students will work very diligently trying to get their own "squawker" to function.

The value of the resistor must lie between 180 and 470 ohms with a value of about 270 ohms being optimum. The capacitor between pins 1 and 4 – 5 determines the pitch of the tone heard in the speaker. Anything close to the value shown will work. The capacitor used to couple the energy to the speaker can be almost any value, with the larger values producing a louder volume. The speaker may be any PM speaker or headphones. If high-impedance headphones are used, the output coupling capacitor may be omitted. The theory of operation of the circuit is given in Chap. 3.

In addition to being a "hooker" circuit to capture interest in making something work in electronics, this first circuit is included in the text for another reason. In a classroom at the beginning of a semester, the instructor has a myriad of things to do. Parts and supplies must be issued. Lockers and textbooks must be issued. A seemingly endless stream of paperwork from the counseling and administration offices must be taken care of. The instructor must handle all this as well as the young and eager students in his charge and retain some semblance of order in the chaos. The circuit of Fig. 1-1 is intended to be a buffer for the instructor. Get the students working with their minds and hands as quickly as possible and take a deep breath as you tackle the paperwork. The necessary groundwork for this experiment has not yet been laid, but you may be surprised at the number of students who will get the circuit operational in spite of that fact. Use the faster students to assist the slower ones.

For the reader at home, give the circuit a try. If you are not successful on your first attempt(s), do not despair. The necessary groundwork will be laid in subsequent chapters, and most of the confusion will be eliminated. The circuit and the experiment may be bypassed by the late enrollees without detrimental effect to the overall educational process.

Chapter 2

Diodes

Refer to Appendix C for the parts list for this group of experiments. Obtain these parts. Components listed for a group of experiments may be used again in other experiments. Once a component has been listed, it is assumed to be available for later experiments and will not be relisted. If a listed component is utilized in some type of support circuitry, however, then a subsequent relisting will be necessary.

The Light Emitting Diode, or LED

A diode is an electronic device that allows electrons to pass through it in one direction only. (Flowing electrons constitute current.) When the diode is connected so that it conducts (electrons flow), it is said to be *forward biased.* When the connections to the diode are reversed and electrons do not flow, the diode is said to be *reversed biased.*

Diodes must never be operated with forward bias without a current-limiting resistor in series with the diode. A forward-biased diode acts like a closed switch (dead short). If the current-limiting resistor is omitted, this short circuit will conduct all the electrons it can possibly conduct and will probably destroy the diode.

A Light Emitting Diode (LED) is a special kind of diode. It gives off light when it is forward biased. The two leads that exit the LED package have names. One lead is called the cathode, and the other lead, the anode.

You will discover that electronics people abbreviate almost everything! LEDs and such are just the beginning. With a little effort on your part, you too can learn to play this game that all the other electronics people play—this game of talking in a foreign language that sounds like English but is totally incomprehensible to the outsider.

On all diodes the cathode end of the diode is identified. On an ordinary diode, the identification will usually be a band near the end of the diode. With the LED, it is usually a flat cut into the material that houses the

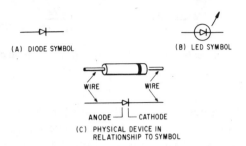

Fig. 2-1 Diodes: (a) conventional symbol, (b) LED symbol, and (c) physical device and relationship to symbol

LED and the leads. However, since errors and exceptions do occur, the ultimate authority on which end is the anode and which the cathode will be you.

The symbol used in electronics to represent a diode is shown in Fig. 2-1(a). The straight lines exiting the symbol represent wires. In this instance, they represent the two wires coming out of the diode itself. (In electronics a straight line represents a wire. The insulation is never shown, but you are all to assume that the wire is insulated unless you are specifically told otherwise.)

The vertical line in the diode symbol represents the cathode. The arrowhead portion of the symbol represents the anode. The current flows in the direction opposite to that of the arrowhead. The symbol for the LED resembles the basic diode symbol but has a circle drawn around the symbol and an arrow pointing away from the symbol, as shown in Fig. 2-1(b). This additional arrow indicates that the LED gives off light when energized. Figure 2-1(c) represents the physical diode in relationship to the schematic symbol.

Resistors

The symbol for the resistor in Fig. 2-2(a) is shown in Fig. 2-2(b). The resistor symbol is a zigzag line between two straight lines. Again, the two straight lines represent the two wires that exit the body of the resistor. Resistors are identified by their color bands. These color bands give the value of the resistor in ohms since the unit of resistance is the ohm (symbolized by the greek letter, Ω). We will learn how to read resistance values shortly; for now, merely orient the resistor so that the color band closest to the end of the body of the resistor is on your left. The first color band will be the first number, the second color band the second number, and the third color band the number of zeroes to be added after the first two numbers. Occasionally, it is very difficult to determine which end of the resistor the colored bands start from. A clue here is that they will always start at the end of the resistor *opposite* the gold or silver band.

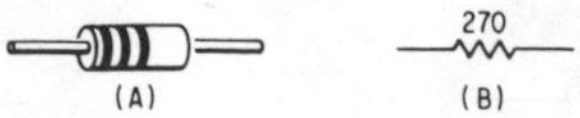

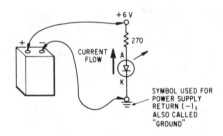

Fig. 2-2 Resistors: (a) physical device, and (b) schematic symbol

Experiment 1

Figure 2-3 illustrates the circuit diagram for the first experiment. The LED is shown connected with forward bias. The resistor shown is a 270-ohm resistor (red-violet-brown). This resistance value is not critical, and any value between about 150 and 1000 ohms can be used. (Note that different values will cause different intensities of illumination.) This resistor must not be omitted, however. If you fail to use a current-limiting resistor, the LED will give a very brief flash before it is destroyed!

Fig. 2-3 Circuit diagram for Experiment 1

This is an extremely simple experiment, but it is amazing how many students manage to get it wrong. To get us started out on the right foot, the following instructions will be unusually detailed.

On both the lantern battery and lab supply, + 6 V usually indicates the positive terminal. Connect a wire from the plus terminal on the lantern battery to the outside rail on the Superstrip. Now take another short jumper wire, and connect the outside rail on one side to the outside rail on the opposite side of the Superstrip. Place very short jumpers at the center sections of both power rails on both sides of the Superstrip. If red wire insulation is available, I suggest that you use it here to serve as a constant reminder that the outside rails are positive. Now connect the negative terminal of the lantern battery to the inner rail on one side of the Superstrip. Use a jumper wire to connect the other inner rail on the opposite side to minus as well. And again, place very short jumpers at the center of the Superstrip to distribute the power down both of its sides. Black makes a good choice of color here to identify the negative power rail.

Place one end of the resistor into any hole along the +6 rail. The other end of the resistor is placed in any five-pin cross-connector. Place one lead of the LED into this same group of five connections. The other LED lead is placed in any *other* 5-pin connector. (Be sure not to get both leads connected into the same group of five connectors; it will be shorted out!) Now take a short length of hook-up wire and jumper the remaining lead of the LED to the ground rail.

The LED will light up if it is forward biased. If it does not light up, it may be reversed biased. Turn the LED around, reversing the connections. If it now lights up, it is forward biased and was reverse biased in the initial connection. If you were successful in getting it connected correctly the first time, then deliberately reverse the connection of the LED in the circuit so that it will be reversed biased and you can verify that it must be forward biased to light. Return it to the forward biased position so that it is lit.

It is conceivable that you may not get the LED to light regardless of the polarity of connection in the circuit. If so, it is probably for one of the following reasons:

1. You could have a bad LED.
2. The LED could be connected incorrectly.
3. Sufficient power is not getting to the LED.

Here we come to the most fundamental point of experimental electronics. In spite of the simplicity of this first circuit, it may not work for any combination of the reasons given above. And the most important lesson of all to be learned is that only *after* an electronic circuit works is anybody in an excellent position to determine why it didn't function in the first place. In this circuit, as well as in any other electronic circuit, we must get it to work first before we can tell what was wrong with it originally!

The fact that the LED does light up tells us three things:

1. We have power applied to the breadboard.
2. The LED is forward biased.
3. The circuit is connected correctly.

To troubleshoot, first try a second LED. *Do not* throw away the first LED. Only after you have an operational circuit with an LED illuminated will you be able to determine if the LED you are about to discard is good or not. If the second LED lights up, then retest the first; if it fails to light now, you may throw it away. If the second LED does not light up either, then the odds are mounting that you have another problem. Doublecheck the connections from the power source to the breadboard. Doublecheck your circuit connections. Doublecheck the color coding on the resistor. Try another resistor. Try another battery or power source; this one might be dead. Use a voltmeter and measure the voltage at each lead of the LED. With a red LED, you should measure about 1.5 to 2.0 V across the LED when it is forward biased. (Appendix D describes the functions and uses of a multimeter.)

The importance of this circuit and its significance to your education is difficult to emphasize on the printed page. It is such a simple circuit that it is almost impossible for it not to work. Right? *Wrong!* In a class of 17 adult students, almost half had trouble! So in spite of its simplicity, don't bypass it. The LED test itself can be used later to troubleshoot inoperable IC circuits.

Summary

1. Since no electronic circuit will function without the correct power applied, the first check is *always* to see if power is being applied to the device under test.
2. Since no electronic circuit will function with even a single wiring error, the second step is to recheck *all* circuit connections against the circuit diagram.
3. Since no device is defective until proven defective, you have to make the circuit functional first in order to test suspected defective parts and prove that they are good or bad. Once proven bad, the defective part may be discarded.

Experiment 2

In this second experiment we are going to add a diode in series with the LED and the resistor of the first experimental circuit. (The first experiment must be operational before proceeding to this one.) This diode is shown added to the first circuit between +6 and the resistor in Fig. 2-4. Actually, it may be added in series anywhere in the circuit for the purpose of this experiment. Placing it elsewhere in the circuit will serve as a check on your comprehension of a series-connected component in a circuit.

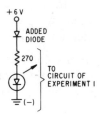

Fig. 2-4 Circuit diagram for Experiment 2

When the diode is added, the LED will remain lit if the new diode is forward biased. If the new diode is reversed biased, no current will flow and the LED will turn off. The diode may be a silicon diode or a germanium diode. Connect it so that the LED stays on; that is, connect it for forward bias. To do so, connect the cathode of the diode toward the more negative voltage in the circuit. (Contrarily, a diode will be reverse biased if its cathode is connected toward the more positive voltage in the circuit.)

A forward-biased silicon diode will have a voltage across it of about 0.6 V. Use a voltmeter to measure the voltage across the diode in the circuit of Fig. 2-4. Place the negative voltmeter lead on the cathode of the diode and the positive voltmeter lead on the anode. A voltage near 0.6 V indicates that the diode is made of silicon. A voltage near 0.2 V indicates that the diode is made of germanium.

This is also a very simple circuit, but one that can reveal several things to you if you keep awake during the experiment. With the LED lit, you can tell which is the anode and which the cathode of any diode. With a voltmeter you can tell if the diode is made of silicon or germanium. And a voltage reading of zero across the diode will reveal that it is shorted.

Experiment 3

In Fig. 2-5 we add still more diodes in series for another experiment. Each added diode will introduce additional voltage drops. Two silicon diodes in series will reduce the voltage available to the LED by 1.2 V. Three diodes would reduce the voltage by 1.8 V. You may even be able to see the brightness of the LED dim as diodes are added. Reversing any of the diodes in the series string will extinguish the LED. A voltmeter used to measure the voltage drops across the diode string can be used to verify that the sum of the voltage drops in a circuit equals the applied voltage.

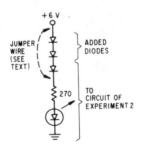

Fig. 2-5 Circuit diagram for Experiment 3

The dotted line in the circuit diagram of Fig. 2-5 represents a jumper that can be placed across the added diodes to take them out of the circuit quickly to see if there is any change in the intensity of the LED with the diodes in and out of the circuit. Since current takes the path of least resistance, it will bypass the diodes and flow through the jumper, causing the LED to light with brighter intensity.

While the circuits and experiments for this introductory session are extremely simple, the educational value of the experiments is difficult to overemphasize. You now know how to test diodes and LEDs, how to determine if you have power to the breadboard, how to determine when a diode is forward biased, how to know which is the cathode and which is the anode on a unmarked (or mismarked) diode, how to determine if a diode is made of silicon or germanium, and how to determine whether the diode is shorted or not. You have also learned the symbols for diodes and resistors and that a straight line is used in circuit diagrams to represent a wire. In other words, you've gotten a foothold on actually translating circuit diagrams into operational circuitry.

For all the rest of the experiments in this book, you must get your test LED circuit operational first. The fact that the LED is lit assures you that you have power to the breadboard. In the LED test circuit of Fig. 2-3, the wire that connects the cathode of the LED to ground will be used to make tests on different components. Observe that a low, or negative, voltage on this wire turns on the LED in the test circuit of Fig. 2-3. *Do not omit the current limiting resistor (the resistor is the component with the colored bands on it) in Fig. 2-3, or you will see a brief flash of light from your LED as you destroy it.*

Chapter 3

Transistors and the Inverter Elements

Modern digital logic integrated circuits (ICs) are made up of transistors, diodes, and resistors. The individual circuit elements are very tiny. They are housed in comparatively large packages so that human beings can work with them more easily.

A diode is manufactured by starting with a pure insulating material such as germanium or silicon and then adding a controlled amount of impurities in a process called *doping.* Doping produces what is called a *semiconductor.* The doping process creates a negative area and a positive area on the silicon or germanium substrate. The boundary between the negative, or n, area and the positive, or p, area is called a *junction.*

A diode has two leads coming out of its package and one pn or np junction. A transistor has three leads coming out of its package and two junctions. Depending on the doping process, these will be either npn or pnp junctions.

Therefore, we have two fundamental types of transistors, the pnp and the npn transistor. Figure 3-1(a) shows the electronic circuit symbol for the pnp transistor; Fig. 3-1(b) shows the npn symbol. Please note that the difference between the two symbols is determined by the direction in which the arrow points. It points toward the base in the pnp transistor and away from the base in the npn. The three leads out of the transistor are called the emitter (E), the base (B), and the collector (C).

Transistors may be conceived of as two diodes back to back inside a package with the base lead as a common connection between the BE diode and the BC diode. Figure 3-1(c) and Fig. 3-1(d) show this internal diode representation for both the pnp and the npn transistor.

Figure 3-2 illustrates common package types for some modern transistors. We are going to do a series of experiments with some transistors, and these may be any transistors that you have on hand. At this time, we do not

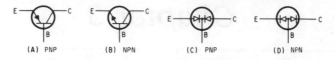

Fig. 3-1 Transistors: (a) pnp, (b) npn, (c) internal representation of pnp, and (d) internal representation of an npn

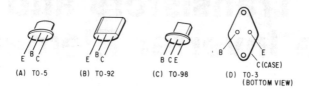

Fig. 3-2 Typical transistor packages: (a) TO-5, (b) TO-92, (c) TO-98, and (d) TO-3

even need to know the 2N number, the type, or the material from which the transistor is made. We can determine all this experimentally, and we can tell whether the transistor is good or bad.

Experiment 1

We start the experiment by setting up the LED test circuit from Chap. 2. We will repeat this set-up procedure several times until it becomes clear that the test circuit must function first before any useful information may be gleaned from subsequent tests. The LED must be operational (lit) first. We then know that we have power and ground applied to the breadboard and that the test LED circuit is functional. Insert a transistor into three different connections on the breadboard. Arrange the transistor so that at least one hole is still visible in the connector and so that you can make connections to the base, emitter, and collector leads. Make sure that you don't plug in the transistor so that two of the leads go into the same group of five; if they do, those two leads will be shorted out.

Start this experiment by connecting the LED and its current limiting resistor as shown in Fig. 3-3. The LED will light when it is forward biased. Get the LED turned "on." Now lift the jumper that connects the LED to ground, and connect it to one of the three leads on the transistor. If you can

+6 V

Fig. 3-3 LED test circuit

identify the base lead on the transistor, connect the jumper to it as shown in Fig. 3-4. If you cannot identify the base lead, then connect the jumper to the middle lead.

With another jumper wire, connect one of the remaining transistor leads to ground. Again, if you can identify the emitter lead, connect it to ground. If you cannot identify the emitter lead, then connect either of the two transistor leads to ground. The object is to turn the LED on again. Experiment. Try switching the two leads around. Try another pair of transistor leads. Once the LED is lit, the diode inside the transistor package is forward biased, and as a result you can identify its anode and cathode. Refer back to Fig. 2-4, and note that the internal diode is pointing in the same direction.

But wait a minute. Suppose the transistor is shorted. Then we do not have a diode inside the package at all! We must reverse the two connections to the two pins that we think comprise a diode to tell if it is shorted. Reversing the connections to the two pins will reverse bias the internal diode, and the LED will go out if we don't have a shorted diode. With one good diode verified, we can now turn our attention to the remaining diode. We again test two of the transistor leads with both forward and reverse bias to verify the direction of the other diode.

Now with two good diodes and their internal configuration determined, we can examine Figs. 3-1(c) and 3-1(d) and determine if the diode configuration is that of an npn or a pnp transistor. You were able to determine which lead of the transistor was a common connection for the diodes. Can you now say with some authority which is the base lead of your transistor? Can you also state that you have a good or bad transistor?

With either diode forward biased inside the transistor package, measure the voltage across this diode with a voltmeter. A voltage of about 0.6 V indicates that the transistor is made of silicon; a voltage of less than half this value indicates that it is made of germanium. Measure the voltage across the LED. Is the LED made of silicon or germanium?

Figs. 3-4 and 3-5 show the experimental set-up for the pnp transistor, and Figs. 3-6 and 3-7 show the experimental set-up for the npn transistor. Figure 3-8 shows a dual LED test set-up. If many transistors need to be tested and their type determined, two LEDs can be used simultaneously to

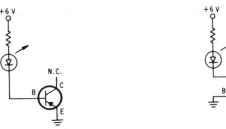

Fig. 3-4 PNP BE junction test   **Fig. 3-5** PNP CB junction test

Fig. 3-6 NPN BE junction test   **Fig. 3-7** NPN CB junction test

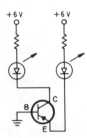

Fig. 3-8 PNP dual-diode test circuit

test both internal diodes and transistor type. Would you need still another dual tester set-up to check the internal diodes for reverse bias?

Now you might have been very lucky and finished this first experiment without any trouble. However, in electronics we have something called "Murphy's Law." Murphy says that "Anything that can go wrong, will." If you were able to bypass Murphy on this initial experiment, don't feel that he isn't around. You will get plenty of opportunity to meet up with Mr. Murphy. In fact, the chances are very high that he has already gotten to you! Go back to Fig. 3-3 and make certain that the LED test circuit is still working. Then return to Fig. 3-4, find one of the internal diodes, and get it forward biased. When you have it forward biased, the LED will turn on. You might just have an npn transistor and if so, then Figs. 3-6, and 3-7 will apply instead of Figs. 3-4, and 3-5.

Between the two test configurations given in Figs. 3-4 through 3-7, we should be able to determine the internal diode configuration of any transistor, and by knowing how the diodes face inside the package, we can determine if we have a pnp or an npn transistor. By using a voltmeter and measuring the voltage drop across a forward-biased diode, we can determine the material from which the transistor was made. By identifying the lead that is common to both internal diodes, we can identify the base lead. Now if we can just find a way to determine either the emitter or the collector, we will have most of the information we need about the transistor.

There are at least two ways to identify the remaining two leads of the transistor, but I am going to postpone this test for the time being. You have enough work ahead of you already at this point in your learning.

For now, buy, or otherwise obtain, a commonly known transistor type. Transistors have what are called 2N numbers. For example, a 2N404 is a pnp germanium transistor in a TO-5 package. The TO-5 package resembles the transistor package depicted in Fig. 3-2(a). A 2N3563 is an npn transistor in a TO-92 package; this package is depicted in Fig. 3-2(b). Use a transistor manual; look up your transistor using its 2N number to provide a known starting point. If you have a power transistor, depicted by the TO-3 package shown in Fig. 3-2(d), then you will need to attach jumper wires to the transistor so that it may be connected to the breadboard for testing.

Turning On Transistors

A transistor is turned on by injecting current (electrons) into the base-emitter junction. When a transistor turns on, it effectively connects the emitter and collector. A small amount of current into the base-emitter diode to forward bias this diode will connect the collector to the emitter inside the transistor. The current that then flows between the emitter and the collector will be limited only by the amount of resistance in series with the collector or the emitter. A small current can thus control a larger current. The small current is in the base circuit, and the larger current is in the emitter-collector circuit.

Experiment 2

Refer again to Fig. 3-3 as the starting point of this experiment. Again we want to make the LED test circuit functional first and verify that we have power to the breadboard and that everything is ready for the experimental work.

Figure 3-9(a) shows the test circuit for the pnp transistor. Note the current-limiting resistor in the base circuit. The value of the resistor may be any value between 470 and 1000 ohms. Remember that the base-emitter junction is a diode and that current must be limited through any diode when it is forward biased to prevent its being destroyed. The wavy lines with an arrowhead on both of their ends represent jumper wires. As shown in Fig. 3-9(a) the jumper may be connected either to ground or to the positive rail (which is + 6 here).

To turn on *any* transistor, you take the base toward the collector. For a pnp transistor, this means that – is applied to the collector; by grounding the base, we turn the transistor on. If the base of the pnp transistor is taken to +6, the transistor will be turned off. With the circuit of Fig. 3-9(a), you can verify this. Taking the base jumper wire to ground will cause the pnp transistor to conduct, thus connecting the collector and emitter, completing

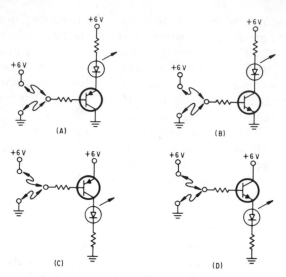

Fig. 3-9 Transistor testing

the circuit, and turning on the LED. Connecting the base jumper wire to +
6 will reverse bias the BE junction, thus turning off the pnp transistor,
disconnecting the collector and emitter, and turning off the LED.

Figure 3-9(b) shows the same experimental circuit, but for the npn
transistor. Since the npn transistor has + applied to its collector, we turn it
on by forward biasing its base-emitter junction by taking the base jumper to
+ 6. Again, when the transistor turns on, the base input current connects
emitter and collector, thus completing the series circuit from ground
through the transistor through the LED and its current-limiting resistor to
+ 6; as a result, the LED turns on. Note how the two different types of
transistors act. A low (any voltage close ᵗ ɔ ground with respect to the
positive voltage) into the pnp turns on the   ED in that circuit, but a high
(any voltage close to the power source witɦ ᵉespect to ground or zero) into
the npn turns on the LED in its circuit. Not that a current in a series circuit
is the same in all parts of the circuit.

What happens if we change things a.ound a little and reverse the
locations of the LED tester in the two circuits. Figures 3-9(c) and 3-9(d)
show this circuit configuration. What do you predict will happen in the
circuit of Fig. 3-9(c)? In Fig. 3-9(d)? A low into the pnp will turn it on. Will
the LED also turn on? What about the npn circuit. A high input to the base
will turn on the npn. Will this turn on the LED? You will have your answers
as soon as you connect up the circuits and try them.

To turn on the LED in the test circuit requires the completion of the
current path from +6 V to ground. In the circuits of Fig. 3-9, the transistors
act as switches to complete the current path from + 6 V to ground. With the

collector of Fig. 3-9(a) grounded, the base of the pnp transistor must be taken toward ground to turn the transistor on. A low (ground) on the base of the pnp transistor switches it on, thus connecting the emitter and collector, pulling the emitter low and completing the circuit, turning on the LED. A low in produces a low out, which is what is needed to turn on the LED in this circuit confirguration.

Please note the following before we proceed:

In Fig. 3-9(a), a low in produced a low out (noninversion)

In Fig. 3-9(b), a high in produced a low out (inversion)

In Fig. 3-9(c), a low in produced a high out (inversion)

In Fig. 3-9(d), a high in produced a high out (noninversion)

Transistors may be connected in circuits in three different basic configurations. All circuit configurations use the terminal that is "grounded" as the reference point. The quotation marks around "grounded" were placed there deliberately. A base, emitter, or a collector is considered "grounded" if no signal is applied or removed from that element of the transistor. The lead may have voltage applied and still be considered "grounded" because it is not concerned with control of the signal through the transistor.

Each circuit configuration has specific properties. These properties are of major concern in the field of analog (linear) electronics but only of minor concern in the field of digital electronics.

The three circuit configurations are: grounded base, grounded emitter, and grounded collector. Figures 3-9(a) and 3-9(b) are examples of the *grounded collector configuration.* This circuit configuration is also called the *emitter follower configuration.* Since the collector of the pnp transistor of Fig. 3-9(a) is directly connected to ground, this circuit is easily seen as a grounded collector circuit. However, in the npn circuit of Fig. 3-9(d), the collector is connected directly to + 6. The collector is "grounded" through the power supply. *The point here is that the collector signal level cannot change.* The voltage at the collector will always be the power supply voltage. No signal can be developed at the collector of the npn transistor in the emitter follower configuration. Note that both grounded collector configurations do not invert the input signal. This is one of the characteristics of the emitter follower configuration.

Figures 3-9(b) and 3-9(c) have the emitter "grounded." The load (the LED test circuit) is placed in the collector circuit in both of these circuits. In both circuits, inversion occurs. A high in produces a low on the collector for the npn circuit, whereas a low in produces a high out for the pnp circuit. One of the characteristics of the common emitter circuits of Figs. 3-9(b) and 3-9(c) is this signal inversion.

Transistors are used to make integrated circuits (ICs). The foregoing experiments were intended to give a bit of background about the nature of transistors. The experiments were far from exhaustive, in fact, just barely sufficient to give you the background to understand what is to follow.

Integrated Circuit Inverters

Refer to Fig. 3-10(a). Four transistors, four resistors, and two diodes are shown connected in a circuit. Four connections to the circuit are also shown: power, ground, input, and output. All the transistors are npn transistors. The input is shown going to the emitter of the transistor on the left. With voltage felt on the base of this transistor, the transistor is enabled and the signal flows to the collector. This is known as the *common base configuration* for transistors. The base of this transistor is returned to + through a resistor in order to forward bias the transistor and turn it on. The diode on the emitter of this transistor is called a "clamp"; it will be reverse biased by any positive signal that is applied to the input. If a negative voltage is applied to the input in excess of 0.6 V, this diode will conduct because it will be forward biased by the applied negative voltage.

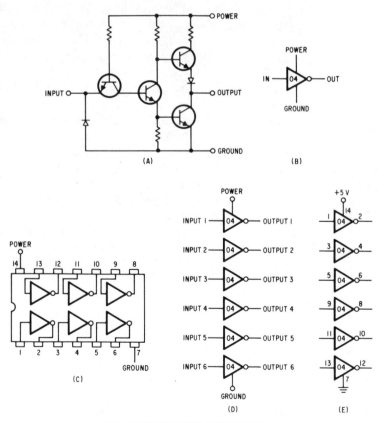

Fig. 3-10 The 7404 hex inverter

A low input to this group of transistors will cause the upper transistor at the far right to turn on and produce a high out. A high in will turn on the lower transistor at the far right, thus connecting the output to ground. The circuit produces inversion and is therefore known as an *inverter.*

There would not be much point in having you connect up the circuit of Fig. 3-10(a) because all the components shown there are to be found in an integrated circuit (IC) called the 7404. In fact, the circuit is repeated six times in this particular IC.

Rather than draw all the components shown in Fig. 3-10(a) every time we wish to indicate an inverter, we use a symbol consisting of a triangle with a small circle that indicates input, output, power, and ground, as shown in Fig. 3-10(b). And since all the chips we will be using are part of the 7400 series, we can omit the "74" and just place the last two or three digits of the particular IC in the center of the symbol to indicate that it is the one we are using.

Figure 3-10(c) shows a top view of the Dual In-Line Plastic Package (DIP) that we will be using throughout this text. Superimposed on the package outline are the six inverter elements inside the package; the pin connections, or pin-outs, are also shown. Note that power and ground are common to all six inverter elements inside the package. Except for the common power and ground connections, all six inverter elements inside the package are independent of each other.

Figure 3-10(d) shows the six inverter elements drawn outside the package. The six inputs are on the left; the six corresponding outs, on the right. If we adopt the standard electronic convention of always showing the inputs on the left and the outputs on the right, we can simplify the IC representation even futher.[1]

Figure 3-10(e) is what may be called a working diagram. It presents all the information given in the data manuals for the pin-outs for the IC, but it is easier to read, interpret, and work with. Since we have already hammered home the point that no circuit will function without power and ground applied, and since most units of the 7400 series have power on pin 14 and ground on pin 7, we will very shortly begin leaving out this information as well.

Some History

The first family of digital logic chips was the 900 series—a resistor-transistor family that operated on 3.2 V. The family was called *RTL,* an acronym for Resistor-Transistor Logic.

RTLs subsequently gave way to another family of digital logic chips

---

[1] It is not always possible to follow this convention in drawing circuit diagrams. However, by the time this may become a problem, your background will be strong enough for you to interpret nonconventional diagrams.

made up of diodes and transistors. This family was called *DTL*, an acronym for Diode-Transistor Logic.

The next family to appear was the Transistor-Transistor Logic family, or *TTL*. After an initial surge of many different numbers, the 7400 series of TTL ICs came out on top in popularity. An old catalog lists a 7400 at $58.00. The price today is under 25¢. Since it is the most widely used (and consequently the least expensive) family available today, it will be the preferred chip in this text.

Other families have since appeared, such as PMOS, NMOS, CMOS, and I²L, but these newer ICs are more expensive than the TTL; the latter still represents the best compromise for experimental purposes. Since other chips all have their place in electronics, we will get to some of them later on.

IC Testing

Experience has shown that an inoperational experimental circuit is rarely the fault of the IC. However, since it is human nature to suspect anything and everything else first when a circuit does not function, each of the experiments that we will be doing will give an IC test circuit first. You may use the test circuit on the IC to verify that it is good before you proceed with the experiment, or you may bypass the test and do the experiment, returning to test the IC if the circuit proves inoperable. The choice is yours, but the test circuit will be there should you need it.

Figure 3-11 shows how to test the 7404. The 7404, called a *hex inverter* in the data manual, has six inverter elements in the same package. To test it, we first set up the LED test circuit and verify that it is operational, as shown in Fig. 3-11(a).

The TTL family is a 5-V family. It will operate reliably on voltages between 4.75 and 5.25 V. The absolute maximum supply is 5.5 V. A silicon diode placed in series with the + 6-V lead from the battery is used to drop the voltage 0.6 V to 5.4 V at the breadboard rails, as shown in Fig. 3-11(b). This diode may be mounted on either end of the wire that runs between + 6 on the lantern battery and the positive rail on the solderless breadboard. It will drop the voltage enough to allow all 7400 circuits to operate normally.

Once the LED tester operates on the new + 5-V source, the resistor connected to the positive rail is moved to any open connection on the breadboard, and a jumper wire is connected to this end of the resistor. Make this jumper wire 6 to 8 inches long to allow room for moving it about the Superstrip.

Place the wire initially on pin 2 of the 7404. The LED should be *out* because TTL inputs that are not connected (left floating) assume a logic high state. If you want to verify this fact, use a voltmeter to measure the voltage on pin 1 of the 7404. It should be + 1.5 V. Since the 7404 interprets this as a "high" on pin 1, the output is low and the LED is off. If pin 1 is

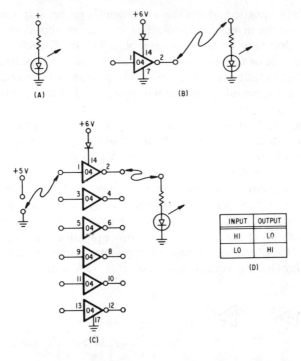

Fig. 3-11 Testing the 7404 hex inverter

now grounded with a jumper wire from pin 1 to the negative rail, the LED will turn on. Since the 7404 is an inverter, the low in produces a high out, thereby turning on the LED.

A logic high for TTL ICs is any voltage greater than 1.5 V. A logic low is any voltage less than 0.8 V. Since any voltage between 0.8 and 1.5 V may be misinterpreted by the TTL IC, operation in this range is "forbidden." (You can operate in this range if you want to, but there is no guarantee as to what the TTL IC will do.) To make it a little easier for you to remember: A high is any voltage near + 5 and a low is any voltage near ground, or 0 V.

Figure 3-11(c) shows the entire 7404 set-up for testing. The jumper wire on the outputs (the jumper wire going to the LED test circuit) is moved from one output to the next while the corresponding input is grounded. With the input Lo, the LED should be on, it will be off if the input is floating or connected to + 5 V.

Figure 3-11(d) shows what is called a *truth table*. A truth table is simply a table in which all IC inputs and outputs are listed. The truth table for the inverter is very simple. A high in produces a low out, whereas a low in produces a high out. This can be stated as follows. The output is not the input.

The "not" portion of the last statement is reflected in the small circle that is part of the inverter symbol. The inverter symbol is shown again in Fig. 3-12(a), but now we have two symbols. The small circle may be drawn on the input side or the output side of the triangle. The indication remains the same: The output is not the input. The position of the small circle also has an additional meaning, but we will postpone discussing this until a little later.

Figure 3-12(b) shows the electronic symbol for the capacitor. It may be drawn in two different ways. Both plates of the capacitor may be straight lines, or one of the plates may be curved. A plus symbol next to one of the plates indicates a *polarized* component. A polarized component is one that is inserted in a circuit in one direction only. Failure to observe correct polarity on capacitances usually will produce a short circuit. The absence of the plus symbol indicates that the device is nonpolarized and may be placed in the circuit with either lead facing the more positive voltage. The numbers next to the capacitor give the capacitance value in microfarads (μF) and the voltage rating in volts (in this case, 10 μF @ 15 V working).

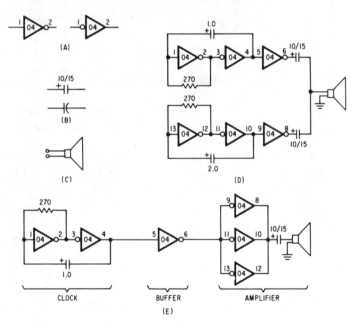

Fig. 3-12 Using the 7404 as a pulse generator

Figure 3-12(c) shows the electronic symbol for a speaker. The two leads exiting the symbol are connections to the voice coil. There exist older dynamic speakers that have more than two connections, but these are quite rare. If the particular speaker application is critical, then additional information will be drawn next to the circuit symbol. The absence of any such information may be interpreted by you to mean that the speaker is not critical and that any speaker may be used.

Let's put the 7404 to work. Figure 3-12(d) shows a rapid 7404 tester if you have many 7404s to test. Each half comprises a tone generator, and both tones will be heard simultaneously in the speaker. The pitch of the tones heard may be varied by changing the values of the 1.0- and 2.0-μF capacitors. Nothing is critical in the circuit except getting your wiring in the right places. The feedback resistors may be any value between 180 and 470 ohms, with the optimum value about 270 ohms. The values of the capacitors may vary widely, but if they are too large, you will not hear tones but pulses, and if they are too small, the tone will be above the range of hearing. The volume of the speakers may be increased by using larger coupling capacitors between pins 6 and 8 and the speaker. The speaker may be any permanent magnet speaker or headphones.

Figure 3-12(e) shows the next experimental circuit to be constructed. The first two 7404 sections form what is called a *clock circuit*. A clock is a pulse generator. The 270-ohm resistor connected from output back to input "confuses" the first section. It does not know whether it is an inverter or an amplifier. The feedback capacitor from the second section to the first provides an in-phase signal to the first section; it is this signal that makes the combination oscillate. The third section acts as a buffer. In electronics, a buffer is any circuit that separates one circuit from another. Here, the buffer separates the clock circuit from the following circuit. The last three sections are connected in parallel to form an amplifier. TTL sections may be connected with the output pins connected under certain conditions. Since the corresponding input pins are also paralleled, we can connect the output pins in parallel as shown in Fig. 3-12(e).

The output of the circuit is coupled to the speaker with a capacitor. The larger this capacitor, the louder will be the sound in the speaker. The speaker may be returned to either ground (as shown) or to plus. To get current to flow through the speaker and produce sound, one of the speaker leads must complete the path through the power supply. The power supply itself offers a very low resistance path (actually a low impedance path) between + and ground. Therefore, the remaining speaker connection may be returned to either + or ground, and the speaker will produce sound. If you return the speaker connection to plus instead of ground, then be sure to reverse the coupling capacitor so that its positive lead points towards the more positive voltage (that is, be sure to observe the polarity of the coupling capacitor).

The speaker is a *transducer*. A transducer is any device that changes one form of energy to another. Here, the speaker changes electrical energy to mechanical energy, which subsequently produces a sound.

The speaker voice coil is a short length of wire. Since it appears as a dead short to the 7404 output sections, a coupling capacitor is included. The circuit will work without a coupling capacitor, but the 7404 is looking at a short circuit. Without the capacitor, it will get hot very quickly and may be destroyed or the speaker may be damaged.

Once the circuit of Fig. 3-12(e) is operational, we can perform an additional experiment or two. First exchange the locations of the 1.0-μF and the 10-μFcapacitors. This will slow the clock down so that pulses will be heard (or a very low tone), and the individual clicks will be audible through the speaker. This experiment illustrates the basic time constant formula: Time Constant= R $\times$ C.

Next, return the circuit to its original configuration so that it produces a tone in the speaker. Now remove the jumpers connecting pins 9 and 13 to pin 6. This step removes two amplifier sections, and the volume out of the speaker decreases significantly. Now add the amplifier sections one at a time, and observe the increase in volume as each section is added. This portion of the experiment illustrates two things. First, that current in parallel is cumulative (adds), and second, that the ear responds logarithmically to the volume of the sound.

The Open Collector 7400s

In Fig. 3-13(a), we have repeated the circuit diagram for a single 7404 inverter section. Observe the two transistors on the right-hand side of the figure. These two transistors form a series path between the positive voltage source and ground. This output configuration of TTL integrated circuits is called a *totem pole configuration.*

In normal operation, one of these two output transistors is "on" and the other is "off." If the lower transistor is turned on, the output line is clamped to ground. If the upper transistor is on, the output line is clamped to plus. If both transistors are on, the only limitation on the current flow through the totem pole pair is the resistor that feeds the collector of the upper transistor in the pair. This is a small resistance in the TTL family; thus if both transistors are on, we have a problem.

Figure 3-13(c) shows two 7404 inverter sections with the outputs tied together. With a high input on the upper section and a low on the lower inverter, we have the same situation as depicted in the previous paragraph. The upper transistor clamps the output line to plus. The lower transistor clamps the output line to ground. This combination makes a *very low* resistance path between the positive power supply and ground, and something has to *give.*

You will usually find in the literature a statement to the effect that ordinary TTL ICs may not be operated with their outputs in parallel as shown in Fig. 3-13(c). Now you have been given the reason why this configuration should be avoided. In Fig. 3-12(e), we operated three inverter sections with their ouputs in parallel but could get away with it because we

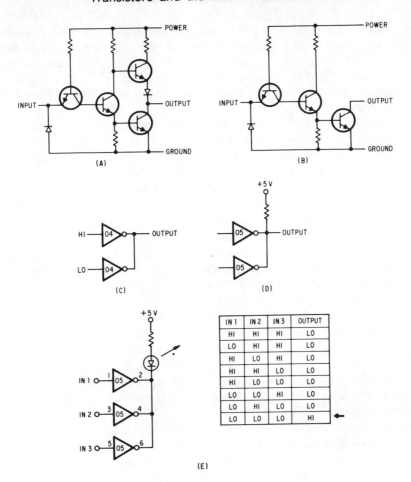

Fig. 3-13 The 7405 open collector hex inverter

operated the *inputs* to the paralleled sections in parallel as well. TTL totem pole outputs may be operated with their outputs in parallel only if their corresponding inputs are operated in parallel.

A group of TTL integrated circuits can also be produced with the upper totem pole transistor, resistor, and internal diode missing from the output configuration. These ICs are called *open collector integrated circuits.* Figure 3-13(b) shows this configuration. Two inverter sections may also be operated with their outputs in parallel without regard to the inputs; this circuit configuration is shown in Fig. 3-13(d). An external resistance must be supplied between the plus voltage source and the paralleled outputs.

In Fig. 3-13(e), we have an experiment for you to perform with an open collector inverter IC, typified here by the 7405. The resistor may be any value between 150 and 1000 ohms for the circuit to perform satisfactorily. The truth table for this experimental configuration is also given in Fig. 3-13(e). Note that with all three inputs floating (they will assume a high if they are left floating), the output is low and the LED turns on. If any combination of inputs are taken low, the output will go high and the LED will turn off.

We have, in effect, created what is called a *gate*. In fact, this is a three-input NOR gate (see pg. 39). With three inputs, 2^3, or eight, possible combinations of highs and lows can be applied to the inputs; consequently, the truth table lists all eight input combinations and their resulting output. Note that only *one* combination of inputs produces a *different* output result. Only if all inputs are low do we get a high out and the LED turned off. For every other input combination, we get a low out, which will turn the LED on.

Problem: Can you arrange the circuit on the breadboard so that the LED will be turned on with a low input and off with all inputs high? Try your hand at designing such a circuit on the solderless breadboard by making use of what you have learned thus far.

The significant line of the truth table in Fig. 3-13(e) is the bottom one. However, the NOR function is actually given by the other seven lines of the truth table. The truth table is really saying, "The output will be a low if either input 1 *OR* input 2 *OR* input 3 is low." The line of the truth table pointed to by the arrow actually gives an AND function: "The output will be high only if input 1 *AND* input 2 *AND* input 3 are all simultaneously low." The "not" portion of the logic comes from the fact that the outputs are inversions of the inputs.

This duality of functions from gates is very troublesome to the beginner. What you will find in the subsequent study of 7400 IC gates is that each gate represents two logic functions rather than just one. The information about the gate may call it a NAND gate. It will perform the NAND function if the inputs are considered to be positive. If the inputs to the NAND gate are active low signals, the gate will be performing the NOR function instead of the NAND function. This duality has given rise to the term "negative logic." We will not study negative logic in this textbook since it tends to confuse newcomers (and some old-timers as well). We will emphasize the fact that gates have dual functions and sometimes will AND the signals and at other times OR the signals that are presented to their inputs. The "clue" to what the gate is actually doing is obtained by observing its inputs and determining whether active highs or active lows are used to control what it does.

If you again refer to Fig. 3-13(e) and the accompanying truth table, you can now see what the preceding paragraph refers to. The top seven lines of the truth table OR the inputs, and the last line ANDs the active low inputs. Since inversion occurs in both cases, we have the NOR function in the upper seven lines and the NAND function in the bottom line. Figure 3-13(e) serves as an introduction to Chap. 4 and the 7400 gates.

The LED test circuit used to test the 7405 sections in this experiment differs from the LED test circuit used to test the 7404 circuit. The wire connected to the cathode of the LED is connected to the output of the circuit, whereas in the 7404 test circuit, the cathode of the LED was left connected to ground, and the wire connected to the 270-ohm resistor was moved as the test lead. A high turns on the LED in the test circuit for the 7404, and a low turns on the LED in the test circuit for the 7405.

Do not let this difference confuse you. Either configuration may be used for the LED test circuit. The second circuit, in which the resistor remains connected to plus while the test lead connected to the cathode of the LED is moved for testing, is the preferred circuit. The reason is that the output structure of TTL ICs will sink more current than the source will. What this means is not important at this stage of your learning. What is important is that you know what voltage level on your test lead from the LED test circuit will turn on the LED. You can quickly determine this each time you use the LED tester by touching the test lead from the circuit to either plus or minus and determining which turns on the LED, a low or a high. This step should be undertaken anyway to make certain that you get *some* information from the tester. That is, make sure that it works before you use it for testing.

Chapter 4

The Simple Gates

The TTL family contains a large number of integrated circuit elements called *gates*. The name derives from the ordinary gate in a fence. When the gate is open, things (people and animals) can pass through the gate. When the gate is closed, traffic through the gate is stopped. When a gate is open in digital electronics, we say that it is *enabled;* when it is closed, we say that it is *disabled.*

Figure 4-1 shows the first member of the 7400 family—the 7400 quad-2 input NAND gate. Figure 4-1(a) shows the internal circuit configuration of one section of the 7400's four internal NAND gates. Compare this circuit with the circuit of Fig. 3-10(a). The basic configuration is almost identical for the inverter element and the NAND gate element. The circuit symbol for the NAND gate is a half circle with a small circle on the output indicating that the NAND gate also inverts the input signal.

Figure 4-1(b) shows the DIP package with the four internal elements superimposed on the package outline; the internal connections to the four NAND gate sections are shown going to their respective pins.

Figure 4-1(c) shows our working drawing and gives the same information as Fig. 4-1(b). The power and ground connections are shown on the first NAND gate section.

Figure 4-1(d) shows a different method of drawing the four NAND gate sections. The Data Manual gives the symbol shown in Figs. 4-1(b) and 4-1(c). In earlier years, this was the only way that the 7400 was depicted. In recent years, the concept that the logic gate should reflect the logic function has been pushed to the forefront. For this reason, we now find the symbol for the 7400 drawn in two different ways, as depicted in Figs. 4-1(c) and 4-1(d). By inserting the last two digits of the 7400 identifier inside the circuit symbol, we can still manage to retain the identification even though different symbols are used.

Figure 4-1(e) gives the test circuit for the 7400. The 7400 has two inputs, and if these inputs are "floating" (disconnected), the single output

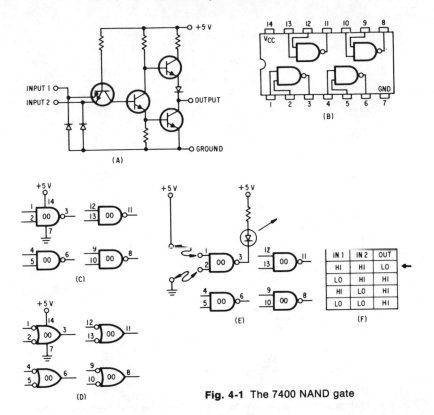

Fig. 4-1 The 7400 NAND gate

will be low. With the LED connected as shown to pin 3 of the first section and pins 14 and 7 connected to power and ground, respectively, the LED will turn on.

If a jumper wire is now connected from one of the input pins to + 5, no change in the LED should take place; it should remain on. If one of the inputs is taken low by placing a jumper wire from that input to ground, the 7400 NAND gate should invert and the output go high. This will turn off the LED. If that jumper wire is left in position connecting one input to ground and a second jumper wire is used to connect the original input also to ground, no change in the LED should occur; it should remain off. If the first jumper is removed and the second left connected, no change should occur in the test LED. Finally, if the second jumper is removed so that both inputs are again high, the LED should again come on. All these conditions are reflected in the 7400 truth table, shown in Fig. 4-1(f).

Note the *true* logic sensing here. *Highs* on the inputs turn the LED *on,* whereas a *low* on one of the inputs turns the LED *off.* Once you understand the truth table and the NAND function, testing all four sections of a 7400 can be quickly achieved. The LED tester lead is moved from pin 3 after the first section is tested and connected to pin 6, which is the output of the

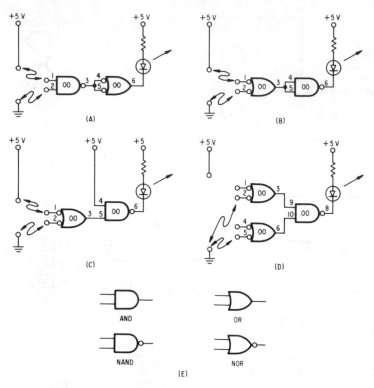

Fig. 4-2 Alternative representations for the 7400 and the four basic logic gates

second section. The corresponding inputs, 4 and 5, are then grounded with a jumper wire, and the LED extinguishes. The LED test wire is then moved to the next output, pin 8, and the two corresponding inputs for this section are grounded. The last section is tested in exactly the same fashion.

At this point, we need to point out a couple of things. First, until you get one section of a 7400 tested, you have no proof that the chip under test is defective. Do not throw away a 7400 (or any other IC that you are testing) until you are able to "prove" your test circuit. Only after you have a test circuit fully operational will you have the necessary assurance that a device under test is good or bad.

Second, never throw away an IC that has more than half of its internal sections functional. Many circuits use only a part of the total gates or elements inside the IC. If two sections of a 7400 are good and two are bad, the IC is still usable. Simply cut off the pins of the defective section (after you *prove* it is defective) so that at some later time you will not end up troubleshooting a problem introduced by the defective section. If more than half of the IC is bad, discard it; it is not worth the effort of saving.

Figures 4-2(a) and 4-2(b) show two 7400 sections cascaded one after the other. Note that the second 7400 section in each case has both input pins connected. When both input pins are connected, the NAND gate section is changed to an inverter section. All it does is invert the signal into it. In both Figs. 4-2(a) and 4-2(b), a low on one of the inputs will turn the LED on whereas a high on both inputs will turn it off. We now have active low inputs where a *low* input turns the LED *on* whereas a *high* input turns it *off*.

The two circuits of Figs. 4-2(a) and 4-2(b) illustrate the concept that circuit diagrams should reflect the logic function performed. The change in the status of the test LED occurs with an active low input. No change occurs with an active high input. Therefore the small circles that indicate active lows should be drawn as shown in Fig. 4-2(b), not as in Fig. 4-2(a).

The two sections should also be drawn differently in each case, with one section drawn as a half circle and the second with this newly introduced circuit symbol that resembles a triply curved circle. The four basic logic symbols are given in Fig. 4-2(e). The AND symbol is the half circle without the small circle, the not AND, or NAND, symbol is the same but with the small circle added to the basic AND symbol. Similarly, the basic NOR symbol differs from the OR symbol only in the addition of the small circle to the basic OR symbol.

Thus, if we have an active low input as shown in Fig. 4-2(b) and the function being performed is the OR function (the change in the circuit output occurs when one input or the other is grounded), we need to draw the circuit as shown in Fig. 4-2(b), not as in Fig. 4-2(a). It should be reasonably obvious that the two circuits will both do exactly the same thing as far as the operation of the LED is concerned. The idea is that the way in which the circuit diagram is drawn should reflect the logic function being performed.

The second 7400 section of Figs. 4-2(a) and 4-2(b) with the two inputs tied together are functioning not as NAND gates but as inverters. Figure 4-2(c) shows another way to use a 7400 section as an inverter. The NAND gate is *enabled* with one input high. In Fig. 4-2(c) you are to tie pin 4 to the positive rail. Observe that the circuit operates exactly the same with this circuit arrangement as it did in Fig. 4-2(b). Now make pin 4 low by jumpering it to the negative rail (ground pin 4). This will *disable* the 7400 section, and the inputs will no longer have any effect on the LED status.

Figure 4-2(d) shows three sections of a 7400 connected in a circuit. As shown, with the inputs floating (they assume a TTL high when floating), the LED is off. Try jumpering pins 1, 2, 4, or 5 to ground. What must you do to the circuit to get a change in the LED? Try different combinations of the input pins (to plus and to minus) until you get a change in the status of the LED. See if you can logically reason out what must be done by making use of what you have learned up to this point. You cannot hurt anything in the circuit by taking inputs to either plus or minus. You can short out your battery if you inadvertently connect the same pin to both plus and minus at the same time. This will not damage the IC, but it will be a little hard on your power supply.

It is considered poor engineering practice to connect the inputs of TTL gates directly to the positive power supply, but we are concerned here not with engineering practice but with education. The inputs to the TTL gates should never go above about 7 V, or permanent damage to the gate input may result. In your experimental work, you should concentrate on what is being advocated here and not be side-tracked. There will be plenty of time later after you get your engineering degree to consider some of the finer points of TTL circuit design.

Additional 7400 Experiments

Figure 4-3 gives some additional 7400 experiments for you to set up on the breadboard and perform. Some circuit symbols for switches are also introduced. Since tying 7400 inputs together makes inverters out of the 7400 sections, we can make a clock circuit similar to the 7404 clock circuit of Fig. 3-12(e). Since this circuit can be drawn in two different ways, both ways are shown in Figs. 4-3(a) and 4-3(b). The capacitor here has been increased to a much larger value so that the circuit will pulse more slowly. This will enable you to see the circuit actually pulsing the LED.

Figure 4-3(c) shows a push-button switch tied to the two inputs of a 7400 gate section. When the switch is pushed, it connects the two associated small circles together and grounds pins 1 and 2 of the 7400. This will turn the LED off. Since the switch is simply connecting these two pins to ground, a jumper wire could be used to accomplish the same thing. The reason for using a switch will be discussed next.

TTL circuits are *fast*. A switch connected to a TTL input to control that input will not make a solid contact when it is closed but will "bounce" several times before the contact is solid. We cannot see the contact bounce with this circuit, but for the moment take our word for it; the bounce is actually there, but it is so fast that we must have additional circuitry to be able to see it.

Control of TTL circuits by using mechanically operated switches is so widespread that circuits have been developed to overcome this switch contact bouncing problem. One method often used is the cross-coupled gate. Figure 4-3(d) shows this circuit drawn correctly—where one push button or the other push button will change the output. Figure 4-3(e) shows the circuit as it used to be drawn for many years; you may still find it drawn in this manner. Set up this circuit on the breadboard and operate it. Since the two push buttons alternately connect pins 1 and 5 to ground, jumper wires may be substituted for the push buttons.

The circuit has two outputs as indicated. The LED tester is connected to one output, and nothing is connected to the second output initially. To

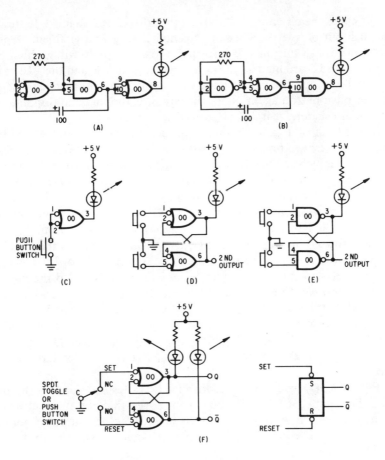

Fig. 4-3 7400 experimental circuits

understand how the circuit functions, you will need all the knowledge that you have accumulated so far.

Assume initially that either pin 2 or pin 4 is high. We will assume that pin 2 is high. Since pin 1 is also high (it is floating and therefore assumes a TTL high), pin 3 must be low. Since pin 3 is connected to pin 4, pin 4 is also low. If pin 4 is low, pin 6 is high regardless of the status of pin 5; thus our initial assumption that pin 2 is high was correct. With pin 3 low, the LED turns on. Now if the top push button switch is depressed, pin 1 will be forced low. When pin 1 is forced low, pin 3 immediately goes high. Since pin 5 is floating and pin 4 goes high because it is connected to pin 3, pin 6 now goes low. Since pin 6 is connected to pin 2, a low is now placed on pin 2, and when the upper push button is released, the output on pin 3 will be held high by the low on pin 2. The circuit is said to be *latched*. The LED is now off with the high on pin 3.

To make the circuit "flip" in the other direction, we must depress the lower push button switch. Since pin 4 is high, this gate is enabled. Depressing the lower push button will force pin 5 low and pin 6 will go high; now both pins 1 and 2 are high so that pin 3 again goes low and the LED again turns on. To "flop" the circuit back over again, we would have to depress the upper push button again. Since the circuit is flipping and flopping back and forth alternately, it is called a flip-flop.

The circuit of Fig. 4-3(f) shows an additional variation of the basic circuit. Here we have replaced the two push button switches with an SPDT switch which may be a push button or a toggle switch. The SPDT stands for Single Pole Double Throw. This switch has three solder connections on it. One solder connection is the "common" (C) terminal. In Fig. 4-3(f), the common terminal is connected to ground. One of the other solder connections on the switch is normally closed to the common connection. When the switch is activated (by depressing the push button or toggling the handle on the toggle switch), the normally closed (NC) contact is opened, and the switch connects to the other solder terminal, which is called the *normally open* (NO) *terminal*. In Fig. 4-3(f), the normally closed contact is connected to pin 1 while the normally open contact is connected to pin 5.

The push button switch depicted in Fig. 4-3(d) is called an SPST, or Single Pole Single Throw, switch. It has only two terminals for solder connection. When this switch is activated, the two solder terminals are connected.

Figure 4-3(f) also has an additional LED tester added to the other output. This added LED will allow you to see that the two outputs are opposite each other; when one output is high, the other output is not high. If one output is called Q, the other output is called $\overline{Q}$. The bar above the Q is read "not-Q".

When a control to a flip-flop makes the Q output go high, it is called the *Set* input. If an input control causes the Q output to go low (or the $\overline{Q}$ to go high, as it is an output which produces signals opposite to those of Q), it is called the *Reset* input. All these conventions are depicted in Fig. 4-3(f). The combination of Set and Reset inputs and Q and $\overline{Q}$ outputs is called a Set-Reset Flip-Flop, or SRFF. The general flip-flop symbol is a rectangle; it is also shown in Fig. 4-3(f).

With the circuit set up on the breadboard as shown, the left LED will be off because the Set input is activated and the Q output is high. The right LED will be on. Activating the switch will cause the left LED to come on and the right LED to turn off. When the switch is released or returned to its original position, the two outputs will again reverse, as indicated by the LEDs switching their status.

Many other experiments with the 7400 are possible, but we will wait until we build up some more background before we tackle them.

The NOR Gate

The 7401 is the next member of the 7400 family. This is a quad 2-input NAND gate like the 7400 but with open collector outputs and different pin-outs than the 7400. The 7403 is a quad 2-input NAND gate with open collector outputs also; the pin-outs for the 7403 are identical to those for the 7400.

The next member of the 7400 family is the 7402. This is a quad 2-input NOR gate. Figures 4-4(a) and 4-4(b) show the IC package and the working drawing. We will dispense with drawing the internal transistor circuitry for additional gates since this information is available in most TTL data manuals.

An examination of the truth table for the 7402 in Fig. 4-4(d) will reveal that the inputs for the NOR gate produce a high output only if both inputs are low. The 7400 NAND gate in Fig. 4-1(e) produced a low output only when inputs 1 and 2 were high. The truth table tells us that to test the 7402

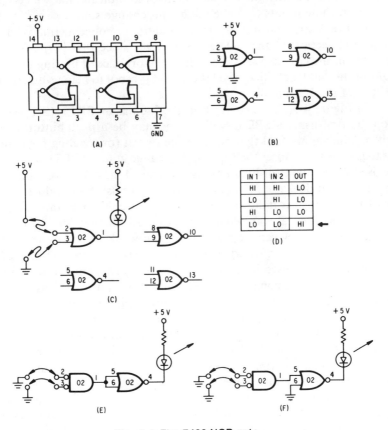

Fig. 4-4 The 7402 NOR gate

we will need to take both the gate inputs low to get a change in the test LED. Connect up the 7402 in the breadboard, and proceed to test all four of its sections to verify the truth table and the fact that all four sections are good.

Figure 4-4(a) shows the 7402 pin-outs, and Fig. 4-4(b), the working diagram. Figure 4-4(c) shows the 7402 test circuit, and Fig. 4-4(d) gives the truth table. Test your 7402 and verify the functionality of all four sections.

A 7402 gate may be operated as an inverter. Figure 4-4(e) shows two 7402 gate sections cascaded with the inputs to the second gate connected in parallel. Figure 4-4(f) shows the alternative connection for inverter operation, with one of the two inputs to the 7402 gate section grounded and the signal applied to the remaining input. Connect both circuits on the breadboard, and verify the operation.

Figure 4-4(e) also introduces another idea. The 7402 is performing the AND function, that is, the output changes only if one input is taken low AND the second input is taken low as well. Therefore, the AND symbol is used in this circuit with the two small circles to indicate that we need active lows in on these inputs to make the circuit change state.

Figure 4-4(f) also introduces an additional concept, that of drawing the second section of the 7402 as an inverter since that is the function it performs in the circuit. We will continue the practice of placing the last two digits of the 7400 identifier inside the circuit symbol to help us determine the identity of the IC actually used.

In Figure 4-5(a), we connect up the 7402 as an SRFF (Set Reset Flip Flop) by first using a SPDT switch. This may be a push-button or toggle switch. You should find that pushing the switch (or toggling the handle on the toggle switch) causes the flip-flop to change state. If SPST push buttons are used for the control function, as shown in Fig. 4-5(b), you should find that the circuit does not function. Depressing a push button does not cause the LED to change as it did earlier in the 7400 circuit. An examination of the truth table will reveal why. *Both* inputs to the 7402 must be low to cause a change of state. How can we make the circuit work with two push buttons?

In Fig. 4-5(c) we have added two resistors (any values between 270 and 1000 ohms may be used here) to pull the "floating" inputs low and enable the 7402. Now if the common connection to both push buttons is returned to plus instead of to ground, the push-button circuit will also work as an SRFF. Set up this circuit on the breadboard, and verify its operation.

This brings us to another experiment. In using a resistor to pull a TTL input low, what range of values of resistance should be used? In Fig. 4-5(d) we have set up an experiment to determine this range. One input to the 7402 is enabled by grounding it (pin 2 is grounded here). A variable resistor is connected between the other input (pin 3) and ground. Since this is our first introduction to the variable resistor, we will have to provide its circuit symbol and the relationship between the symbol and the physical device.

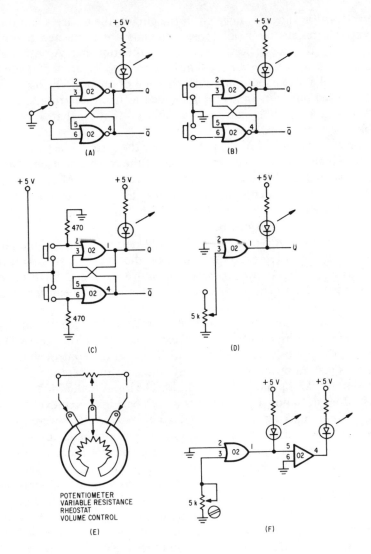

Fig. 4-5 Experimental circuits for the 7402

Figure 4-5(e) is an attempt to convey this information via the printed page. The variable resistor goes by a number of aliases. If the physical device has only two solder terminals, it is known as a *variable resistor* or *rheostat.* If the physical device has three solder terminals, it is known as a *potentiometer,* which is usually shortened to *pot* in electronics literature. If only two of the solder terminals on the pot are connected in a circuit, the device operates as a variable resistance, or rheostat, rather than as a pot. If two of the terminals are shown connected in the circuit, the device again operates as a variable resistance. If all three terminals are connected to different points in a circuit, the device operates as a potentiometer. A potentiometer is also called a *volume control* or, simply, a *control.* Figure 4-5(e) shows the circuit symbol above a pictorial representation of the pot. The basic resistor symbol is used but with an arrow appended to it. In electronics, an arrow included with a symbol indicates variability—that is, that a device is a variable resistance, capacitance, inductor, etc.

The center terminal of the pot is always the arrow. The two remaining solder terminals connect to opposite ends of the resistance itself. The value of the resistance can be changed so that different values of controls can be manufactured. The potentiometer used for this experiment must have a value of at least 2500 ohms. A 5000-ohm pot is a reasonable value, but anything over 2500 ohms can be used.

The circle in the center of the pictorial representation of the physical device in Fig. 4-5(e) represents the shaft on the pot. It is this shaft that is rotated to cause the arrow (or wiper) to move back and forth inside the potentiometer case to change the resistance between one end terminal and the wiper.

Set up the circuit of Fig. 4-5(d) on the breadboard. Rotate the shaft on the control, and observe that at one extreme of rotation the LED is on whereas at the opposite extreme it is off. Now set the control shaft so that the LED is just on or just off. You may find a point where the LED is just a little dimmer. This point is the threshold. The input to the 7402 is such that the 7402 does not know if the input is a high or a low. Remove the control from the circuit, making careful note of which two terminals were connected between ground and the 7402 input, and use an ohmmeter to measure the value of resistance. You should find that it is somewhere near 2400 ohms. However, since both ohmmeters and ICs vary, you may get a value quite different from 2400 ohms. Almost all TTL inputs will interpret a resistance of 1000 ohms or less on that input to ground as a "low" on that input. Thus, to pull a TTL input low with a resistor, use a value of 1000 ohms or less.

Figure 4-5(f) shows an alteration of the circuit of Fig. 4-5(d). An inverter section is added. This allows you to see the threshold more easily by having two LEDs. Additionally, we have connected the potentiometer with two of its terminals shorted together to make a variable resistor. And we have introduced still another electronic symbol in Fig. 4-5(f). A circle with two lines through it next to a variable resistor symbol means *screwdriver adjust.* Potentiometers and rheostats are made in two basic types—front-panel controls that have a fairly long shaft and screwdriver-adjust controls that have short shafts (or no shaft at all) that are intended to be adjusted only occasionally. Either type of control may be used for the experiment by soldering wires to the terminals and plugging these wires into the breadboard.

The 7408 AND Gate

Figure 4-6 gives the introductory material for the 7408 IC. This is a quad 2-input AND gate. The truth table reveals that both inputs to the AND gate must be high to get a high out. Set up the 7408 on the breadboard, and verify the truth table [Fig. 4-6(d)] and all four sections of the 7408.

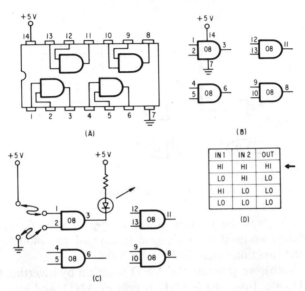

Fig. 4-6 The 7408 AND gate

The 7432 OR Gate

Figure 4-7 gives the necessary information for you to set up and test the 7432 IC on the breadboard. This is a quad 2-input OR gate. To test it, first examine the truth table and see what the inputs to do change the output. Then test the 7432 to verify that it performs as specified in the truth table [Fig. 4-7(d)] and that all four 7432 sections are operational.

Making the Four Functions from One IC

We have examined and studied the truth tables of the four basic logic functions: NAND, NOR, AND, and OR. We have also seen that these four functions are available to us in different ICs, and we have learned how to make inverters from each of the various functions.

It is possible to generate any one of the four logic functions from any one of the quad-2 input gates. We will set up the NAND gate to generate all four functions and leave the generation of all four functions from the NOR gate as an exercise for you.

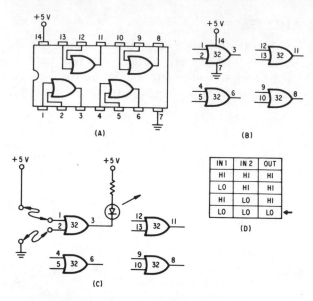

Fig. 4-7 The 7432 OR gate

Figure 4-8 gives the starting point in Fig. 4-8(a). To generate the NAND function, we need use only one section of the 7400 NAND gate since this is the function "built in" to the 7400.

In Fig. 4-8(b), we generate the AND function by inverting the output of the NAND gate. Inverting NAND produces AND, and inverting AND

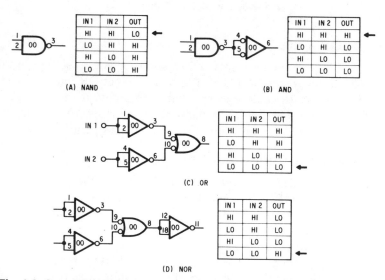

Fig. 4-8 Generating all four basic logic functions using one quad 2 input gate

produces NAND. To produce the OR function, we invert the inputs to the NAND, as shown in Fig. 4-8(c). Finally, to produce the NOR, we invert the OR function with another inverter, as shown in Fig. 4-8(d).

Truth tables accompany each of the circuit functions and can be verified by setting up the functions on the breadboard. Normally, you would buy the IC that gives you the function desired. However, since there are times when spare sections of ICs are "left over" in a circuit design, economy of manufacture dictates that fewer components be used to produce a given device. This is why it is important for you to learn these interrelationships so that if you see the circuits in some other design or in a repair job, you can identify what is happening and troubleshoot accordingly.

Summary

To generate the four basic logic functions—AND, OR, NAND, and NOR—four TTL ICs are available. The remaining logic function is that of inversion, which is performed by the 7404 TTL IC. All four basic logic functions may also be generated with a single quad-2 input gate.

TTL ICs are made with totem pole outputs. These outputs may not be connected in parallel except in certain instances. To connect TTL outputs in parallel, the open collector version of the basic function is used.

We can now test all basic TTL ICs on the breadboard. We can also now summarize the logic functions as follows:

Inversion—The output is not the input.
NAND—All high inputs produce a low output.
AND—All high inputs produce a high output.
NOR—All low inputs produce a high output.
OR—Any high input produces a high output.

Chapter 5

The Flip-Flop

As you can see, it does not take very long for even an introductory text in digital electronics to get "deep."

A flip-flop, or, as it will be called here henceforth, an FF, is an electronic circuit that can store digital information. It is a memory device that can store a high, or a one, and a low, or a zero. These two digits—1s and $\emptyset$s—are the only two in a numbering system called the *binary number system*. You have grown up with a numbering system called the *decimal system*. Both these numbering systems will count to infinity, but ten symbols —$\emptyset$, 1, 2, 3, 4, 5, 6, 7, 8, and 9—are required for the decimal system and only two—$\emptyset$ and 1—for the binary system. We also use the terms "low" and "not true ($\overline{\text{true}}$)" for the symbol $\emptyset$ and "high" or "true" for the symbol 1. Although this can be very confusing at the start, things will soon fall into place.

Chapter 4 introduced the SRFF, which is formed by cross-coupling a pair of gates, and also introduced the basic FF symbol, which is a rectangle. If you are even the least bit hazy about the logic involved in the cross-coupled gate, it is mandatory that you review its operation at this time.

Figure 5-1 shows the cross-coupled NAND gate (a) and the equivalent FF symbol (b). The SRFF has two inputs, called the *SET* and *RESET inputs*. The SET input is also called a *PRESET input* and the RESET input is also called a *CLEAR input*. SET and PRESET are the same, and RESET

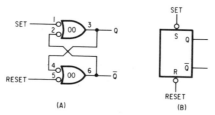

(A)   (B)

Fig. 5-1 The Set/Reset flip-flop

46

and CLEAR are the same. The outputs are called Q and $\overline{Q}$; if one is high, the other must be low.

Small circles associated with the logic symbols for the ICs indicate that the input or output is an active low. Small circles are also appended to basic logic symbols to indicate the *NOT* function. The basic symbol for the AND gate is shown in Fig. 5-2(a). If we add a small circle to the AND symbol, we change the symbol to the NOT-AND symbol. The word "NOT-AND" is then contracted to produce the word "NAND," which is the name given to the symbol in Fig. 5-2(b).

(A) AND   (B) NAND   (C) RESET

Fig. 5-2 The NOT symbol

The small circles associated with the FF rectangle mean that the inputs to the circuit cause the outputs to change when the inputs are taken low. The SRFF of Fig. 5-1 will SET (cause the Q output to take on a 1, or high level) when the input is taken low. The RESET input to the SRFF will cause the Q output to assume a $\emptyset$, or low, logic level. Note that all inputs are defined with relation to the Q output. Since the $\overline{Q}$ will always be the logical reverse ("logical inverse" is a better term) of the Q output, SET causes this output to go low whereas RESET causes it to go high. If you memorize the relationship for the Q output and then remember that the $\overline{Q}$ output is the inverse of the Q output, you will not get quite so confused.

The JK Flip-Flop

There are a great many different FFs available in the TTL logic family. Let us look first at a very common one, the 7473. The 7473 is the first chip in our survey that does not have power applied to pin 14 and ground applied to pin 7. The great majority of the 7400 series has power applied to the "middle" and highest pin, that is, 7 and 14 or 8 and 16, or 12 and 24. There are some ICs, however, that have power applied to other pins. This is just one of the reasons for always connecting power and ground first when an IC is first inserted into a solderless breadboard. Figure 5-3 shows the pin-outs for the 7473.

This IC has four inputs and two outputs for each of its circuits. The inputs are clock, J, K, and CLEAR. We will use the symbol for phase, ϕ, to identify the clock input. We will also ignore the J and the K inputs

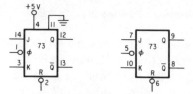

Fig. 5-3 The JK flip-flop

for the time being, but remember that TTL inputs left unconnected, or "floating," assume a TTL logic high.

Since a small circle is associated with the clear input, we know that by taking this input low (grounding it) we can force the Q output low. Note that this FF does not have a SET (or PRESET) input. Also note that it is a dual chip since there are two FFs inside the same IC package. Both sections are independent of each other except for the power and ground connections that are common to both FFs.

Testing the 7473

First connect pin 4 to + 5 volts and pin 11 to ground. Next make up two LED testers and hang one on the Q output of the first FF section and the second on its $\overline{Q}$ output (see Fig. 5-4). One test LED will be off, and the other, on. If neither LED is on, then check the LED tester and the power

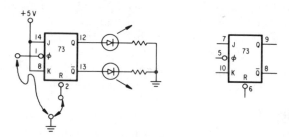

Fig. 5-4 Testing the 7473

connection to pin 4. If both LEDs are on, check the ground connection to the IC. If the test LED connected to the Q output is on, the clear function may be tested by grounding the CLEAR input to the FF. If the Q output is low, as evidenced by the test LED connected to it being off, then testing of the CLEAR function will have to wait for a moment. Use a jumper wire, and momentarily touch the clock input (ϕ) to ground. What we want to do is to get the Q output high (get the test LED connected to the Q output on) so that we can test the CLEAR input. You may find that the "bounce" produced by grounding the ϕ input makes it just a little bit difficult to get the

Q output high. Once you get the test LED connected to the Q output on, you can then ground the CLEAR input to turn it off.

This simple experiment illustrates what was pointed out to you in Chap. 4. Mechanical switch contacts bounce, and TTL ICs will trigger on the bounces. By adding the switch debouncer of Chap. 4 to the circuit, we can get things under control. Figure 5-5 indicates this addition. The cross-coupled NAND gate debounces the "switches" so that we can get only one pulse at a time into the 7473 clock circuit. The "switches" need not be actual switches. Their function is to close an input to the SRFF to ground so that jumper wires may be used to simulate the action of the switches.

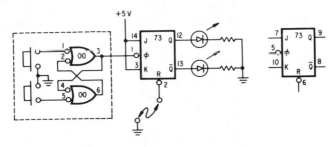

Fig. 5-5 Clocking the 7473 with the SRFF

Now you can check the 7473 under "experimenter control." Clock the ϕ input of the 7473 with the switch debouncer circuit. Each time you clock the ϕ input, the Q output should change state. If it was high, it should go low. If it was low, it should go high. When you have the Q output high, stop clocking the ϕ input; use the CLEAR input to turn off the LED connected to the Q output. This will check the CLEAR function. The other FF in the IC package is tested in exactly the same way.

The J and K inputs must be high for this IC to flip-flop. Try grounding both J and K inputs and repeating the test procedures. The outputs should not change as the ϕ input is pulsed. The CLEAR input should still cause the Q to clear, or go low, but this may only be seen if this output is high when you ground the J and K inputs.

Pull-Ups

For experimental purposes, unused TTL inputs, such as the J and K inputs to the 7473, may be left floating. TTL inputs left disconnected assume a TTL logic high. In all "finished" production circuits, you will find these unused inputs tied high, usually to the positive power supply through a resistor, 1000 ohms being a common value.

Also for experimental purposes, inputs may be tied high by connecting them directly to the + 5 rail, a practice not usually followed in commercial circuits. Try connecting the J and K inputs with jumpers to the 7473 to + 5 volts, and repeat the test procedures. With the J and K inputs tied high, the 7473 should still "check out," just as it did with these inputs left floating.

If you see resistors tied to unused (and sometimes even used) inputs in circuit diagrams, you will now realize that their function is to make certain that the unused input is kept at a TTL logic high so that extraneous noise pulses cannot get into the IC to foul up circuit operation.

A Counting Circuit

Figure 5-6 shows how we can use the two sections of the 7473 to perform a counting function. Connect up the 7473 on the breadboard as shown. The Q output of the first FF is connected to the φ input of the second FF. The two test LEDS are connected to the $\overline{Q}$ outputs of both FF sections. The CLEAR inputs of both sections are conncected in parallel.

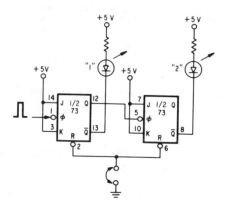

Fig. 5-6 Counting with the 7473

Ground the CLEAR inputs momentarily with a jumper wire; then remove this wire. Both LEDs should now be out. If they are not both out, check to see if you have one, or both, connected to the Q outputs instead of the $\overline{Q}$ outputs.

Now, with the debouncer circuit to provide pulses, input a pulse to the φ section of the first FF. The source of these pulses will be pin 3 of the 7400 debouncer circuit shown in Fig. 5-5 (inside the broken-line rectangle). Remember that a pulse will have to go from high to low and back to high again (or from low to high and back to low again) to make a complete pulse. For it to do so will take two pushes of the "switches" in the debounce circuit (or two alternative closures to ground on the SET and RESET inputs to the debouncer circuit if you are not using switches) to generate a single pulse.

After one pulse is generated, the LED connected to the first Q output should be on. We now have a "one" counted. Generate another pulse with the input circuitry. The first LED should go out, and the second LED should come on. The first LED has a value assigned to it of "one," and the second LED, a value of "two." With two pulses entered, the LED with the "two" value should be on.

Generate another pulse. What do you expect to happen? The LED with a "one" value should come on, whereas that with a "two" value should remain on. Since we have three pulses entered, the values indicated by the test LEDs should reflect the three pulses. A "two" plus a "one" is a "three."

Now what do you expect to happen when the next pulse is entered? Both LEDs are now on, indicating that three pulses have been entered. Generate yet another pulse into the counter to see. All LEDs now go off. We have counted four pulses even though the LEDs indicate that we are back where we started!

That fourth pulse is stored only in your head. You saw the three pulses indicated and watched the circuit return to its original state with the fourth pulse. To have that fourth pulse indicated by a LED's being on, we would need yet another LED to "hold" the count for us. We can continue counting input pulses, and each fourth pulse will be stored only in our heads. But what's this? We are counting in the decimal system, but the circuit is operating in binary, in which only two digits are indicated by the FFs—1's and $\emptyset$'s. How is this possible? We alone make the conversion from one numbering system to the other. By assigning a decimal value to the *position* of the LED in a circuit, we are able to indicate all values in the system with which we are more familiar.

Each FF will count and display (with the LEDs) a total count of 2. Each added FF will increase the count ability by a power of 2. Two FFs will count to 2^2, or 4. Three FFs will count to 2^3, or 8. Four will count to 2^4, or 16. Each added FF will continue to double the counting ability, and the chain of FFs can be extended indefinitely—2^5 will produce 32, 2^6 will produce 64, and so on. The ability to count to infinity in the binary system differs from that in the decimal system only in that two digits are used instead of the ten in the decimal system (nine numerals *plus* zero).

Connect up the second 7473 now with all the CLEAR lines to each FF connected in parallel. The ϕ input to each succeeding FF is connected to the Q output of the preceding section. Four LED test circuits connect to the four $\overline{Q}$ outputs (see Fig. 5-7).

By entering 16 pulses into the first ϕ input with the debouncer circuit, we can now count to 16 and have the count combinations stored for us by the LEDs. The first LED represents a 1, the second a 2, the third a 4, and the fourth an 8. As each LED is illuminated, the cumulative count is obtained in the decimal system by adding together the values displayed by the LEDs. When the LEDs are positioned so that the LED displaying the 1 is the closest to you on the breadboard, and the LED displaying the 8 is the farthest from you, they are arranged in what is known as the standard "vertical format" of binary display. If the entire console and/or breadboard

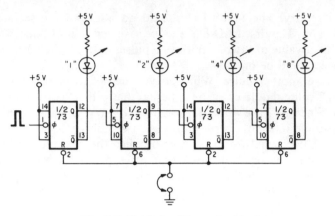

Fig. 5-7 Divide-by-16 counter block

is then rotated 90 degrees counterclockwise so that the Least Significant Digit (LSD) is on the right and the Most Significant Digit (MSD) is on the left, the LED presentation will then be the standard "horizontal format" of binary display.

With the added FF sections, we can now count to 16 before we must store the count in our heads. Additional FF sections and LED indicators may be added to count and display any value we desire.

Gating

Since we are much more familiar with the decimal system, it would be more convenient if we could have the electronics do some of the decoding for us and thus simplify the conversion from the binary to the decimal system. Let's see what is involved.

The "shift" in the decimal system occurs after we have used the numerals $\emptyset$ through 9. To indicate the next count value in the decimal system, we use what is called *positional notation*. In positional notation, the number 1 has a value of 1 when its position is 01. If the position of the 1 is changed, it takes on a different value. The numbers 01 and 10 are not the same and do not represent the same count even though the same two symbols are used. When a 1 is shown followed by a 0, we call this number a *ten*. Likewise, the value of the 1 changes if we position it yet elsewhere, as in 100. Here the count represented is no longer one or ten but *one hundred*.

Of course, you knew all this already, so what is the point? The point is that we must use four FF sections to count to 10 because three FFs would count only to 8. And we will have to trick the FFs into counting in the decimal system because they normally count in the binary system. To do so, we will have to use some gating.

At the count of nine, the LED display indicates 1001, where a 1 indicates that the LED at that position is on, whereas a 0 indicates that the

LED at that position is off. The LED presentation of 1001 is equivalent to a cumulative count of 9 since the "8" and the "1" LEDs are on.

On the next pulse that we enter, the display will read 1010. But we don't want the display to read 10 (an "8" plus a "2"). We want it to generate a "carry" into the next column of the "positions" and have the display read 0000. This would then give us the "1" stored in our heads, and the 0000 presented by the LEDs would be correct for a decimal 10.

A NAND gate has a low output only if its inputs are high. If one section of a 7400 is connected to the Q outputs at the "2" and "8" positions in our counter chain, the 7400 will output a low pulse when it sees both its inputs high. This low pulse will be connected to the four CLEAR inputs to the FF chain to reset or clear all four flip-flops to zero. The reset pulse so generated can be fed to still another FF section, and this pulse will activate that section to store the "carry." Figure 5-8 shows you how to gate four 7473 FF sections to make them count to 9 and then read a 0000 (zero) on the next count and generate a "carry" for the 1 in the 10 position.

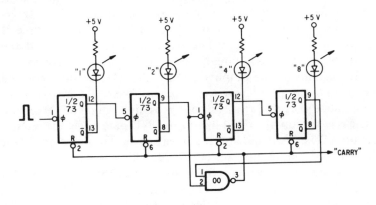

Fig. 5-8 Decade counting with flip-flops

By suitable gating, the counter chain may be made to count to different numbers and then reset and start over again. How can you make this circuit divide (or count) by 6? By 9? By 5?

To make the circuit count (or divide) by 7 (see Fig. 5-9) requires that three inputs be sensed for highs simultaneously. To do so requires a three-input NAND gate such as the 7410. To make the circuit divide by 7, the Q outputs corresponding to "1," "2," and "4" must be gated to reset the FF chain to 0 on the 7 count.

The Digital Die

The singular of the word "dice" is "die." To make a digital die, we must have a circuit that will count to 6 and then display that number. The

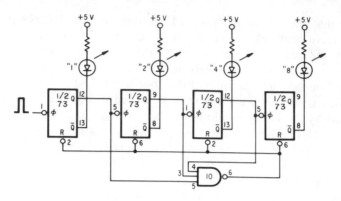

Fig. 5-9 Divide by 7

gating just described will count to 7, but the 7 will never be displayed because the circuit is reset to 0 on the seventh count. Since a zero is never displayed by a die, some additional gating must be used to eliminate it from the die. One way to gate out the zero is to use another 7410 three-input NAND gate and gate the $\overline{Q}$ outputs. On the count of zero, all three of these outputs are simultaneously high. The problem now is that we need to tie two TTL outputs together even though we learned in Chap. 4 that to do so is a no-no. We cannot tie the outputs of both the 7410 sections together and connect them to the clear line of the FF chain. We could use two sections of a three-input AND gate (whose output goes high only if all three inputs are high) and one section of a quad-2 input NAND gate to get the required low to reset the 7473s. We could also use inverters on the 7410 outputs and then use the inverted outputs to gate one 7400 section to accomplish the same thing. There are also other ways to eliminate the zero from the display. The gating required for this will be left as an exercise for you. Please remember that there are several solutions to this particular problem. *Any* solution that you can devise will get you an "A" for your efforts.

The 7474

Another type of FF is the D flip-flop, and a widely available TTL chip is the 7474. Figure 5-10 shows the pin-outs for the 7474. This chip has power on pins 14 and 7. It is also a dual, with two D FF sections in the same package, and has a PRESET (SET) and a CLEAR (RESET) pin. Both of these inputs are active lows, indicated by the small circles on the FF symbol. Again, both FF sections are independent except for the common power and ground connections.

The D describing the flip-flop type stands for "Delay." It is easier to understand how this chip functions if we reassign the function for the D input and call it the *data input*. Whatever is on the D input at the time the FF is clocked will be transferred to the Q output. Likewise, since the $\overline{Q}$

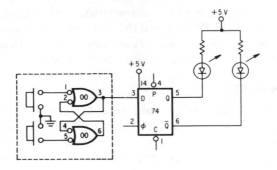

Fig. 5-10 The 7474 Dual D flip-flop

output is the inverse of the Q output, the complement of the signal on the
D input will appear at the $\overline{Q}$ output.

To test the 7474, connect LED testers to the Q and $\overline{Q}$ outputs. Reset
the FF by momentarily grounding the CLEAR (RESET) line. Now SET the
Q output by momentarily grounding the PRESET (SET) input. In the first
case, the Q output will go low; in the second case, it will go high.

Next CLEAR the FF and clock the ϕ input. (Use the debounced pulse
generator in the broken-line rectangle of Fig. 5-5 to generate the clock
pulses.) When the D input is floating, it assumes a TTL high; this high
should be transferred to the Q output. Subsequent clock pulses should not
change the Q output; it should remain high. Now ground the D input with
a jumper wire. Again clock the D Flip-Flop. Now the low on the D input
should be transferred to the Q output, and this output should remain low
with continued clock pulse inputs. Repeat the tests for the second D FF
section.

Now connect the $\overline{Q}$ output back to the D input, and clear the flip-flop.
Now the $\overline{Q}$ output is high and the Q output, low. The high from the $\overline{Q}$
output is applied to the D input. When we clock the FF, the high on the D
input will be transferred to the Q output, and the latter will go high. The $\overline{Q}$
output will go low, and now a low is applied to the D input. On the next
clock pulse, the low on the D input will be transferred to the Q output so
that the D FF will act like the 7473 (when used with both J and K open).

The output of one FF can be connected to the ϕ input of the second
section to make a counter chain again. 7473s and 7474s can be combined to
make a longer counter chain.

As we shall see later, the ability to transfer the data on the D input to
the Q output of a flip-flop will come in extremely handy.

Other Flip-Flops

There is a wide variety of flip-flops available in the TTL logic family.
We have JK FFs, D FFs, J$\overline{K}$ FFs, FFs with negative-edge transition

clocks, FFs with positive-edge transition clocks, FFs with only the RESET input, FFs with both the RESET and the SET inputs, and so on. All the flip-flops store 1s and 0s. All are basically memory devices. All can be used for counters. They are indeed a *basic digital building block*.

Troubleshooting

All necessary material for your succesful completion of the experiments in Chapter 5 has been included. In spite of this, too high a percentage of students are unable to make them all work. The basic problem turns out to be one of assuming that everything taught has been learned. In the real world of the classroom or at home, this assumption is unwarranted. What we really need here is a back-to-basics review of what has been presented up to this point.

1. You must assume that an IC is good until you prove that it is bad. It is human nature to place the blame on the IC rather than on the individual who connected it. When it is possible that an IC is bad, put it back in the test circuit and retest it. Students will occasionally destroy an IC with incorrect connections, but this is a rare occurrence.

2. You must always check your circuit connections and make certain that they are correct. However, you will be able to *know* that they are correct only after the circuit works. At any prior time when the circuit is inoperative, you must consider the remote possibility that you have made an error in connecting it.

3. Reread the text. It is just possible that you are attempting the impossible because you have misinterpreted what the circuit is supposed to do.

4. Try to break up the entire circuit into smaller sections, and test each of the circuits separately. In Fig. 5-9, for example, there are several subsections. Test the SRFF to make sure that you have a clock pulse out of it. Test the pulse at the output and at the input to the clock input for the first 7473 section. Test each LED readout circuit separately. Does each one function? Get the first counting stage to work first; *then* add the next stage after the first one is operable. Progress down the counting chain one FF at a time, getting each successive FF to function. Now start adding the gating circuits one lead at a time, and test at each addition.

5. There is nothing so maddening as an inoperable electronic circuit. After you get the circuit operational, you become an instant expert. You know exactly what was wrong with it. Until the circuit is operable, no one is an expert, not even your instructor. The "mag ic" that he performs to get your circuit functioning isn't magic at

all. He is just as baffled initially as to why your circuit won't work as you are. He depends on his years of experience to know what to look for and his years of troubleshooting discipline to track down the problem. You are in the process of learning this discipline now. There is always a reason for a circuit's not functioning. Make like a detective. Look for "clues." Make logical deductions based on what your "clues" are trying to tell you.

6. Back to basics! Pull the chip and examine the pins. Did you bend one when inserting it in the breadboard? Do you have power and ground on the breadboard? Is the polarity of power and ground correct? Do you have power and ground at the pins of the IC itself? No, not on the breadboard, but right where power and ground enter the chip. (This is the only way in which you will discover whether pin 14 or 7, or in the case of the 7473, pin 4 or 11, is bent.) Do you have the reset line grounded? Is it supposed to be high or low? Check each pin in numerical order, starting with pin 1. Above all, do not give up. Stay with it until you whip the problem. Sometimes when one cannot find out why a circuit does not work, it is a good idea to tear it down and start all over. This approach is fine for experimental work but cannot be used later on the job.

Chapter 6

Pulse Generation

At some time or other during your experimentation, the recreating of auxiliary circuits on the breadboard each time they are needed will become a pain. The continued skidding around of the solderless breadboard will also become very annoying. Moreover, the voltage of the lantern battery may approach the point where replacement of the power source is called for.

Chapters that will help you solve these problems will be found toward the end of the book. The chapter on power supplies includes a supply suitable for operating the breadboard. The chapter on consoles solves some of the additional problems. Whenever you feel that some improvements in performing the experiments are highly desirable, "branch" to the chapters in question and see what they can do to help you out.

Clocks

We have previously defined a clock for digital systems. A clock is a circuit that generates pulses. The cross-coupled NAND gate that we have been using allows us to go from a logic low to a logic high and back again by pushing a switch. This creates pulses and hence serves as a clock circuit. To be sure, this is an extremely slow clock, but there are many times when we need a clock that will run only as fast as will allow us to see what is happening.

The 7404 clock generator encountered in Chap. 3 was also a simple clock. That circuit generated pulses at an audio rate so that a tone was generated when the pulses were coupled to a speaker. We also used the same circuit later to generate slow clock pulses to pulse an LED. If we are to see the LED flash, the pulse rate must be less than about 16 Pulses Per Second (PPS) because of the phenomenon called "persistence of vision." The human eye tends to retain images for about one-sixteenth of a second; any images presented to the eye at a faster rate will seem to be present to the mind continuously.

The entire world of digital electronics operates on clock pulses. There are innumerable ways to generate them. To be able to see them we can use a slow clock and pulse a LED. In faster clock circuits, we can "see" them by listening with an audio transducer. A transducer is any device that changes one form of energy to another form. In this case, we change electrical energy from the clock circuit into audio energy by using a speaker as a transducer.

To see pulses faster than the pulses generated within the audio range (roughly, 20 to 20,000 Hz), we must use an oscilloscope. An oscilloscope, or scope, is an electronic device that changes electrical energy into light energy. It is a transducer, but you will rarely hear anyone refer to a scope as such. The internal operation of an oscilloscope is quite beyond the range of this text. Most educational facilities today have a scope available. Therefore, we will assume that an oscilloscope is available to you, as well as an instructor who can show you how to use it. To see pulses, we need a scope. Don't rush out and buy one, but find a resource person so that you can see the pulses on his scope.

An oscilloscope for digital work is a fairly expensive item. It is not something ordinarily found in the inventory of the electronics enthusiast. Amateur radio operators, or hams, will often have one available. You will find the majority of hams very friendly and more than willing to assist you in any way that they can. If you don't know any hams, all you have to do is to drive down a street and look for the ham's giveaway antenna farm. Park your car, ring the doorbell, and introduce yourself. You'll make a new friend almost instantly and one who may be a very valuable resource. If this particular ham does not have a scope, he is very likely to know one who does. And he can always get "on the air" and in a few minutes find one. Who knows, you might even get "hooked" on amateur radio in the process, or equally likely, get the ham hooked on digital electronics.

The starting place for this experiment is to reconnect the earlier 7404 clock circuit on the breadboard and get it operational. You can tell if it is working by the tone in the speaker. This operating circuit can then be "visualized" by observing the pulses it produces on the scope.

Figure 6-1 shows the scope output for the 7404 clock generator. Refer back to Fig. 1-1(b) for the actual breadboard circuit using the 7404.

The 7413

This clock generator uses a member of the 7400 family that has hysteresis built into the circuitry of the chip. The word *hysteresis* means that the turn-on point and turn-off point for the digital circuitry are not the same. The 7413 resembles the 7420 in its circuit pin-out configuration. Figure 6-2(a) shows the working diagram for the 7420, Fig. 6-2(b) shows the working diagram for the 7413, and Fig. 6-2(c) shows a test circuit for both. This test circuit is a go-no go test and does not show the hysteresis. To show the hysteresis, we need an additional test circuit as well.

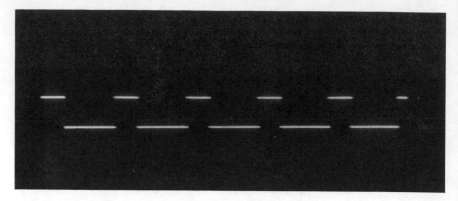

Fig. 6-1 Scope output for the 7404 clock generator

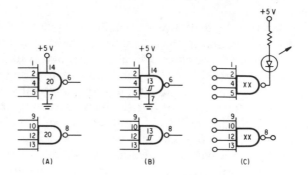

Fig. 6-2 The 7420 and the 7413

The 7420 pin outs are identical to the pin-outs for the 7413. Up to this point we have not really experimented with the multiple input gates. The 7420 and the 7413 are both dual 4-input gates. Their truth tables would reveal that their outputs are low only if all inputs are high. The initial test for these two gates, then, is to hang the LED tester on an output and observe if the LED is on. (Don't forget to have power and ground hooked up first!)

Grounding any one of the four inputs will turn off the LED. Grounding any additional inputs after the first is grounded will keep the LED off. Test all inputs and test both sections of the chip. Note that on both these chips pin 3 and pin 11 have no connection internally. Do not discard a chip with only one defective input. Clip off that pin and keep the gate for that circuit that does not require all the inputs to be used. Discard any chip that has a defective output for one of the two sections (discard all chips that are only half good or less).

Figure 6-3 is a very simple circuit that shows the hysteresis. The first

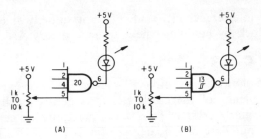

Fig. 6-3 Hysteresis

step in the experiment is to use the 7420 for a comparison. Connect it up first, as shown in Fig. 6-3(a). As the control is turned, you should be able to find a point where the LED dims. This is the point where the input does not know whether it is high or low; consequently, it alternately turns the output off and on. The point is a fairly critical one and is much more easily seen with a ten-turn pot. A ten-turn pot is a control that requires ten turns of the shaft to traverse from one end of the resistance value to the other. An ordinary control will suffice for this experiment if you have suitable patience.

Now replace the 7420 with the 7413, as shown in Fig. 6-3(b). Again rotate the control shaft. You should find that you cannot dim the LED. It is either full on or full off. You should also find that the LED turns on at a different shaft setting than it turns off. A voltmeter can be added to the circuit to measure the two hysteresis points if you should desire to do so. A digital voltmeter will prove more useful, but the analog vacuum tube voltmeter (VTVM) or its solid state equivalent are usable on the lowest range.

Figure 6-4 shows the experimental 7413 clock circuit. Since only half the 7413 is used as the generator, the other half is operated as an inverter/buffer for isolation. The control pot here will most likely be limited to 1000 ohms or less. In the author's own experiments, the feedback resistance had to be less than 1000 ohms or the circuit stopped functioning. The timing capacitors in the circuit can be changed, and the clock generator will run from essentially dc to about 25 MHz. This is an extremely stable RC circuit and can be used for a wide number of applications. Since it will operate through the entire lower radio spectrum, it will generate con-

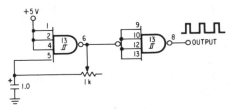

Fig. 6-4 The 7413 variable clock

siderable radio interference (RFI) if it is not suitably shielded. You can listen to this clock by holding it next to an ordinary AM radio receiver.

The Crystal Clock

Figure 6-5 shows the "standard" 7404 crystal clock generator. In this circuit, the frequency of the clock pulses is controlled by the resonant frequency of a quartz crystal. This is not a required experiment, since it necessitates the purchase of a quartz crystal. It is included here as a reference for crystal controlled clocks should you need such a circuit in the future. If a crystal is readily available, by all means try the circuit, but don't make a special purchase at this time.

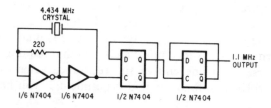

Fig. 6-5 The standard crystal-controlled clock (*Courtesy*, Signetics Corp.)

Figure 6-6 shows the versatility of the earlier 7404 clock circuit. By replacing the feedback capacitor with a quartz crystal, the circuit will generate clock pulses at the crystal frequency. Since this circuit is simpler than the "standard" circuit of Fig. 6-5, we prefer to use it whenever a crystal controlled clock is called for.

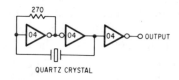

Fig. 6-6 The versatile 7404 clock circuit used as a crystal-controlled clock

The outputs of both circuits may be checked with a frequency counter if one is available, and they may also be viewed on the scope.

The 555 Clock Generator

The 555 chip is not a TTL chip; it is TTL compatible. This means that the output of the circuit will drive the input circuits of TTL chips; it also means that the 555 will operate on + 5 V. The 555 is classified as a linear chip. It is a timer and may be used as a clock generator for TTL circuits. Because of its versatility, it may also be used for a large number of other

TTL applications. Its frequent use as a clock generator is the reason for its inclusion in this chapter on pulse generation.

Figure 6-7 gives the pin-outs and the test circuit for the 555. This is our first non-TTL chip (our first linear chip) as well as our first eight-pin chip. Exercise a little caution. Power is on pin 8 (not ground), and ground is on pin 1. Linear ICs blow quickly if power is inadvertently reversed. The output of the 555 is adjusted by the control to provide pulses in the audio range; thus we can hear the pulses generated by the circuit. Since this audio checker is the fastest means of testing 555s that we have found, it forms the test circuit as well. This test circuit will not test all the input functions for the 555, but it does provide a test that lets you evaluate a 555 on a go-no-go basis. Additional testing is needed for the VCO input (pin 5) and the reset pin (pin 4). The 556 is a 14-pin DIP version of the 555; it has two 555s inside the same package.

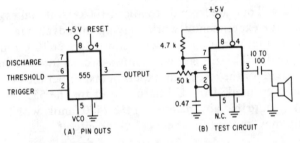

Fig. 6-7 The 555 and the 555 clock

Figure 6-8 is a slightly modified version of Fig. 6-7. We might dub this circuit the "squawker." Most of the time an ohmmeter is used not to measure ohms, but continuity. The "squawker" is an audio continuity tester. If the circuit under investigation has the required continuity, the "squawker" sounds off with a tone or squawk. Its advantage over the ohmmeter is that you need not take your eyes off the circuit you are testing to determine the continuity.

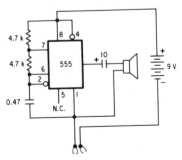

Fig. 6-8 The "squawker": An audio-continuity tester

The 555 operates over a voltage range of 4.5 to 18 V. The power supply for the "squawker," therefore, is most conveniently provided by a 9-V transistor radio battery. A piece of lamp cord will provide the leads, and a pair of alligator clips attached to the lamp cord are a convenient way to connect the "squawker." Since the clip leads switch the battery power to the chip on and off, no off-on switch is needed.

The speaker may be any small speaker salvaged from a defunct transistor radio. Any small enclosure may be used to house the test instrument. In fact, one student came up with a very practical solution. He took the "guts" out of a transistor radio, leaving the speaker and battery intact. He then added the "squawker" circuitry and had the enclosure, the battery, and the speaker problem all solved. Once he added a small PC board (discussed in Chapter 6), test leads, and clips, he had a small, compact, rugged test instrument in jig time. Don't fight it; build yourself a "squawker." You will wonder how you ever got by without it.

The 555 will operate from dc to about 100kHz. It makes a satisfactory clock circuit for experimental work, having an advantage over the 7413 clock circuit in that it will operate with controls over 1000 ohms. The 7413 operates beyond 100 kHz. Figure 6-9 shows circuits for experimental clocks that may be used for the rest of the experiments. Either the 7413 clock or the 555 clock may be used. The PC board layout for the 555 clock is given in Chap. 16. Also note that any member of the 7400 family with hysteresis built into the chip may be used for the 7413 clock circuit.

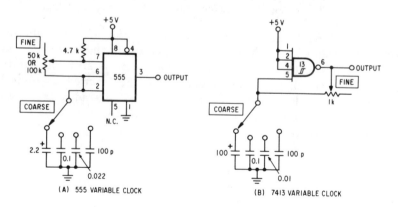

Fig. 6-9 Two variable clocks suitable for experimental work

The 74121 Family

A group of chips in the 7400 family are "cussed" by engineers more often than they are praised. These are the 74121, the 74122, and the 74123, as shown in Figs. 6-10(a), 6-10(b), and 6-10(c). These chips are called monostable multivibrators. The 74121 is a single nonretriggerable multi-

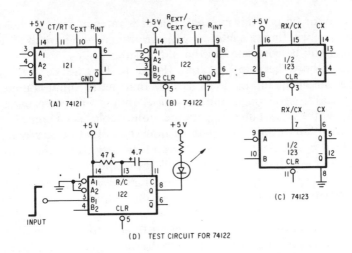

Fig. 6-10 The 74121 monostable multivibrator family

vibrator; the 74122, a single retriggerable multivibrator; and the 74123, a dual retriggerable multivibrator.

The reason that they are cussed so much by engineers is that they are partly TTL and partly linear. Since they are linear, they are susceptible to noise triggering, and this is what gives them their bad reputation. The literature is full of admonishments to "design around" these chips if at all possible.

Yet, since these chips are used, and used often, we need to investigate them. Let us try the 74122, although any, or all three, could be used in our experiments. Refer to Fig. 6-10(b). This chip has five inputs: A_1, A_2, B_1, B_2, and a clear input. The A inputs are active low inputs, that is, the chip will trigger on the high-to-low transistion of the clock input pulse if it is applied to these pins. Either, or both, A inputs may be used. If the A inputs are used, the B inputs may be left floating or tied high. The B inputs are triggers for the low-to-high edge of the input clock pulse. If the B inputs are used, the A inputs must be grounded. If both the A and the B inputs are used in the same circuit, the A input must normally be low and the B input must normally be high for the two of them to function.

The 74122 has two outputs, Q and $\overline{Q}$, with the same meaning as that assigned to the outputs of the flip-flops. The Q output will be the output cleared low when the clear input (or reset input) is activated. The small circle associated with the clear input means that this input is activated by a low applied to it and that for normal operation, it should be at a TTL high.

Some pins on the 74122 are neither inputs nor outputs. These are the timing inputs. R_{ext}/C_{ext} is the pin on which the resistor and capacitor are joined if these two components are external. C_{ext} is the pin to which the "other end" of the capacitor will connect. R_{int} is the end of the internal resistor used for timing; this internal resistor is rated at around 2000 ohms. Normally, you will find this internal resistor unused in the circuit. The other end of R_{ext} will be connected to the + 5 V supply.

Here is how the chip operates. [Refer to Fig. 6-10(d).] A pulse is applied to one of the inputs. Let's assume that the B_1 input is used. Since this is a positive-edge trigger input, both A inputs will be grounded. The rising edge of the clock pulse triggers the monostable. The Q output will go high and then fall again. The length of time the output pulse remains high is determined by the size of the resistor and the capacitor connected to the timing connections. If the values are small, the output width of the pulse will be narrow. If they are large, the width will be wide. A very narrow, positive input pulse will then produce a negative-going output pulse and a positive-going output pulse; their pulse width will be determined by the values of resistance and capacitance. Since the monostable is triggered on the edge of the input clock, the input pulse can even be very wide (that is, the output from the cross-coupled NAND gates), but the output pulse width will still be determined by the RC combination in the timing circuit. Thus, regardless of the width of the input pulse, the output pulse will always have the same width.

The 74122 is retriggerable, which means that the inputs are always active, that they are not "locked out." If a series of input pulses is applied to the clock inputs, the monostable will retrigger, and the Q output will remain high as long as you keep triggering the inputs. Therefore, it is possible to use this chip to generate extremely wide pulses as well as fixed narrow ones. There are times when these different capabilities can come in very handy.

As an example, let's use the 74122 as a pulse stretcher. Narrow pulses can be seen only with an oscilloscope. If we can stretch the narrow pulse, we can observe it with a LED. But first we must generate a very narrow pulse.

Figure 6-11 shows a circuit called the *cheap-shot*. This is a one-shot circuit made from a 7400 inverter section or a gate section. It is a very handy circuit. The resistor on the input holds the input low (at least, the 7404 thinks that this input is low!). This means that the output of this section is high. If a transition from high to low and back to high again is applied to the input (we can generate this transition with the cross-coupled NAND gate circuit), the cheap-shot will be triggered. The resistor and capacitor have been selected to produce a pulse that is about ½ millisecond wide, far

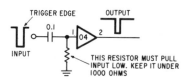

Fig. 6-11 The "Cheap-Shot"

too narrow to be seen with a LED (but easily visible on a scope). Thus, the LED connected to the output will appear unchanged. The output pulse from the cheap-shot is therefore a negative-going ½-millisecond pulse. To get a 1/2-millisecond-wide positive-going pulse we run it through an inverter section (see Fig. 6-12).

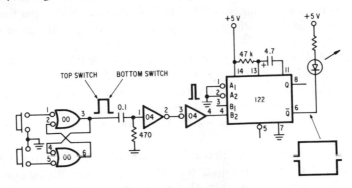

Fig. 6-12 Pulse stretching

This positive-going pulse is fed to the 74122 circuit, and the timing components are selected to give an output pulse width wide enough to allow the eye to see the LED flash. We have now created a pulse-stretcher circuit. Each time we trigger our 7404 cheap-shot and generate a pulse, the 74122 stretches this pulse out and makes it wide enough for us to see with a LED. Increasing the value of either the resistor *or* the capacitor in the timing circuit will widen the output pulse and keep the LED off (or on) longer.

To see the retriggerable operation of the 74122, we need to connect up a clock circuit. Use either the 7413 or the 555 variable clock circuit, and feed its output to the cheap-shot input (see Fig. 6-13). Start with the clock running slowly. The LED should flash as the 74122 is triggered by the cheap-shot. Now gradually increase the clock frequency. As the number of pulses per second (PPS) is increased, a point will be reached at which the

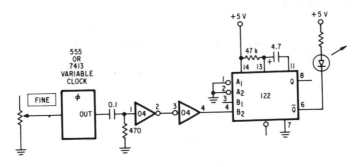

Fig. 6-13 Retriggering

timing values do not "time out" and the 74122 remains continously retriggered. The LED will then remain on (if it is connected to the Q output) or off (if it is connected to the $\overline{Q}$ output). The input clock pulses are arriving too rapidly, and the 74122 is retriggering. The Q and $\overline{Q}$ outputs remain "fixed."

Now reduce the number of PPS by lowering the clock input frequency. When you again reach the point at which the timing values time out, the LED will again flash.

The 74121 is a nonretriggerable monostable. Inside the 74121 chip are circuits that "lock out" the inputs once the 74121 has been triggered. The 74121 must time out before the input circuits again become active. There are times when this feature is needed and you do not want the retriggerable operation. At such times, the 74121 should be used instead of a 74122 or 74123.

If the inputs of the retriggerable monostables (the 74122 or the 74123) are connected back to an input, the circuit will make still another clock generator. Figure 6-14 illustrates this clock circuit, which is to be found in the Signetics Data Manual and uses the 74123. Its output is adjustable around the 1-MHz frequency commonly used for microprocessors.

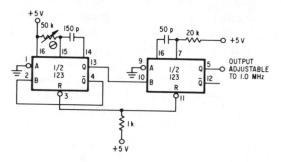

Fig. 6-14 The 74123 clock

Square Waves

The square wave, which is characterized by having equal on and off times, is frequently needed. Its high pulse width is the same as its low pulse width. The easiest way to generate one is to send a clock signal through a flip-flop, as illustrated in Fig. 6-15. Remember, however, that the flip-flop

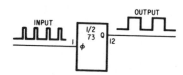

Fig. 6-15 Generating square waves

divides the input clock frequency by two. If you want 1-MHz frequency out, then the input to the flip-flop must be 2 MHz.

Summary

A great deal of information has been jammed into this chapter. At this point, let us try to tie up a few loose ends. In Fig. 6-2 we introduced the dual four-input NAND gate, the 7420. We sneaked it in without much emphasis. We also introduced the 7413 and mentioned hysteresis but did not emphasize the Schmidt trigger symbol (T) associated with the hysteresis built into the chip. In Fig. 6-2(c), we inserted a pair of XXs in the NAND gate symbol and didn't mention that fact either. When digits are unimportant, or variable, they are usually replaced by Xs. Here the XXs mean that the test circuit applies equally to the 7420 and the 7413.

In Fig. 6-4, we introduced the *pulse train,* the string of pulses out of the 7413 clock generator shown at the right. The train depicts graphically what the signal out of the circuit looks like when observed on a scope.

In Fig. 6-5, we introduced the quartz crystal and its electronic symbol. Take note of this symbol for future reference.

In Fig. 6-8, we introduced the electronic circuit symbol for the battery. We had mentioned this symbol in an earlier chapter, but it is time to refresh our memories.

In Fig. 6-9, we introduced the circuit symbol for the rotary switch. As the control knob connected to the switch shaft is turned, the arrow moves from one switch contact to an adjacent contact. This switch provides a convenient means of changing the value of the timing capacitors in the circuit. Since the values of capacitance switched into the circuit will cause the clock frequency to "jump," this control is labeled "Coarse"; the resistance control is labeled "Fine."

In Fig. 6-9, we introduced the concept of *capacitance* as used in timing circuits, but we have not as yet defined this property. Capacitance is the ability to store *electrons,* those constituents of an atom with a negative electrical charge. The unit of capacitance is the *farad.* The more farads a capacitor is rated for, the more electrons it can store.

Since the farad proved to be too large a unit for most electronic work, a metric prefix was added to it to produce a fractional unit called the *microfarad,* abbreviated as μF. The metric prefix "micro" means one-millionth; a microfarad is thus one one-millionth of a farad.

In certain instances, the microfarad itself is too large a unit; hence the *picofarad* was introduced. The picofarad is one one-millionth of a microfarad. Its abbreviation is pF. Since almost all values of capacitance are given in microfarads, the common practice is to omit the unit if it *is* microfarads and indicate it only if it is *not* microfarads. Because electronic nuts tend to be lazy, the "F" of "pF" is very often omitted as well and just the small "p" used to indicate picofarads. This is illustrated in Fig. 6-9, where values of the timing capacitors are simply listed as digits alongside the capacitor symbol. Here, 2.2 means 2.2 microfarads. On the other hand, the 100p means that this capacitor is rated at 100 picofarads.

In Fig. 6-10, we introduced the symbol for "edge trigger." Only that portion of the input clock pulse that triggers the circuit is shown. The symbol starts out low and goes high, indicating that the B input is a positive-going edge trigger. The negative-going portion of the clock pulse does not affect the 74122 input.

In Fig. 6-11, we indicate that the cheap-shot triggers on the rising edge of the input pulse. The entire pulse, however, is drawn for the input clock. This means that the 7404 cheap-shot needs the entire input pulse to generate an output but that the output will be generated on the rising edge of the input pulse.

In Fig. 6-13, still another shorthand convention is introduced. Here the rectangle encloses the clock symbol. The notation means that you can use either the 555 clock circuit *or* the 7413 clock circuit. The portion of the selected clock circuit that is to be varied is the fine frequency adjust; the coarse frequency adjust remains in a single position.

In Fig. 6-15, only the connections needed to understand the circuit function are given. By this time, you should realize that the 7473 shown will not function unless you connect power (pin 4) and ground (pin 11).

How many of these got by you the first time through?

This chapter is extremely important because the entire world of digital electronics and computers is based on pulses. The ability to control these pulses means the difference between an operable and inoperable machine. We have seen how pulses may be generated and their shapes changed, inverted, lengthened, shortened, and so on. Make every effort to get to a scope so that you can see these pulses.

Chapter 7

The Counters

Among the most interesting members of the TTL logic family are the counters; the 7400 family has quite a variety. In Chap. 5, we learned that four flip-flop stages will count to decimal 16. To count to decimal 10, we had to add some gating. Let us pick up the counting game again at this point.

Figure 7-1 shows four cascaded 7473 sections that will count to decimal 15 and on the count of 16 start all over again from zero. To make

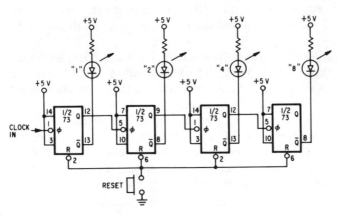

Fig. 7-1 Counting in binary with flip-flops

the four stages count to decimal 10, we need to have the counter reset to zero on the count of 9 and generate a reset or carry pulse. This requirement is again illustrated in Fig. 7-2, in which the Q outputs of the second and fourth stages are fed to a 7400 NAND gate and the output of this gate causes the counter chain to reset to zero. The LEDs monitoring the outputs will count from zero through 9, but on the tenth count, the two inputs to the NAND gate will go high, thus allowing its output to go low and thereby resetting the counters back to zero.

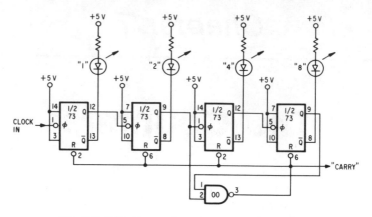

Fig. 7-2 Decade counting with flip-flops and gates

The 7490 Family

By building the gating into the chip instead of using external gates, the IC manufacturers have provided ICs that will divide by 2, 5, 6, and 8. To accomplish the same results that we did with the four 7473 counter stages, we can use the 7490. This IC has two sections—a divide-by-2 section and a divide-by-5 section. Division by 10 may be accomplished in two ways. You can either divide by 2 first and then by 5, or you can divide by 5 first and then by 2.

Figure 7-3 shows both these schemes. Figure 7-3(a) shows division by 2 first and then by 5, and Fig. 7-3(b) shows division by 5 first and then by 2. Figure 7-3(a) is used for conventional counting circuits and for feeding the different decoders to drive seven-segment displays and various digital displays. Figure 7-3(b) is used whenever you want division by 10 but need a square wave output from the last stage. Figure 7-3(a) produces what are termed *Binary Coded Decimal (BCD) outputs,* whereas Fig. 7-3(b) produces what are termed *biquinary outputs.* With biquinary outputs, the "8," "4," "2," and "1" designations to the LEDs no longer have their previous meaning. For example, the "1" output (pin 12, flip-flop A) changes at the slowest, rather than the fastest, rate.

Figure 7-4 illustrates the experimental set-up for the 7492 member of this family. This chip has a divide-by-2 and a divide-by-6 section. The combination allows division by 12, a capability needed in clocks of the timekeeping variety.

Figure 7-5 uses the 7493 member of the family. This binary counter has a divide-by-2 and a divide-by-8 section. One advantage of the 7493 over the 7473 circuits is that the total division may be performed using one chip instead of two.

A couple of points need to be mentioned with respect to this family. First, power is on pin 5 and ground is on pin 10, a totally different

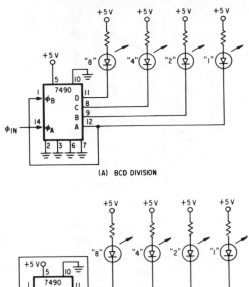

(A) BCD DIVISION

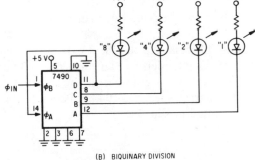

(B) BIQUINARY DIVISION

Fig. 7-3 The 7490 decade counter

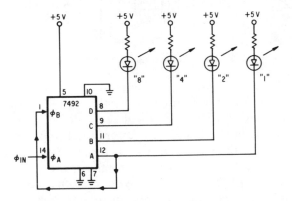

Fig. 7-4 The 7492 divide-by-12 counter

arrangement from that for other chips investigated so far. Second, the reset
lines for these chips are active highs instead of active lows. Thus, if you leave
the reset pins for these chips floating, they won't work. To have each chip
count, the reset lines must all be held low, or at least one reset line in each
group (R_0 and R_9) must be held low. To reset the 7490, the R_0 lines are *both*

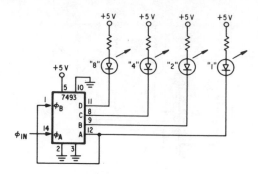

Fig. 7-5 The 7493 binary counter

taken high. The reset lines can be controlled for experiments by jumpers on the breadboard.

By providing multiple reset inputs, the 7490 may be made to divide by other numbers than it was intended to. To have the 7490 divide by 6, for example, the 2 and 4 outputs are gated with the internal gates inside the chip. All that is needed is to connect pins 9 and 8 back to the two R_0 inputs (pins 2 and 3), as illustrated in Fig. 7-6.

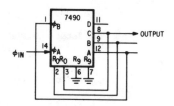

Fig. 7-6 The 7490 used as a divide-by-6 counter

Figure 7-7 shows how to use four 7490s to make a decimal counting chain that will count to 10,000 (9999 will be displayed, and the display will read 0000 on the count of 10,000). LEDs are used to provide the read-out information. This is the type of read-out display used on older pieces of electronic equipment. To get the total count, the operator adds up the values of each illuminated LED (each 7490 provides another decade).

Figure 7-8 shows how the 7492 may be used to make a digital clock for timekeeping purposes. Modern digital clocks use a clock chip in which all counters, gating, and sometimes even the digital display are built into the chip itself.

We will not breadboard the circuits of Figs. 7-7 and 7-8. They have been included here simply for their educational value so that you can see how things used to be done. These circuits are obsolete and are not worth

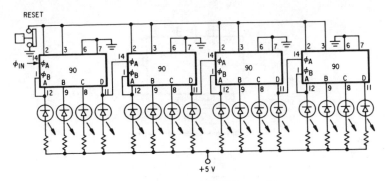

Fig. 7-7 Divide by 1000 counter

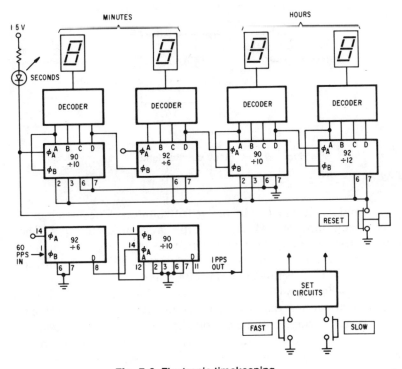

Fig. 7-8 Electronic timekeeping

the cost to construct them as more modern chips do the entire job for less money.

The 74160 Family

The 74160 family is another group of counter chips (see Fig. 7-9). Its members include the 74160 and 74162 decade (divide-by-10) counters and

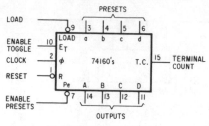

(A) PIN CONNECTIONS FOR FAMILY

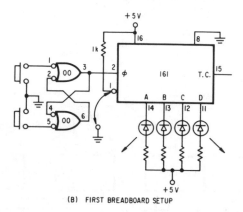

(B) FIRST BREADBOARD SETUP

Fig. 7-9 The 74160 counter family

the 74161 and 74163 binary (divide-by-16) counters. The 74160 and 74161 form one subgroup; the 74162 and 74163, a second subgroup. The subgroups are determined by the way in which the reset circuits function in the chips. These counters are synchronous and presettable. This family group is not divided into two sections like the 7490 family.

In the 7490 family, the count progresses down the chain of flip-flops, one flip-flop triggering the next. The count "ripples" down the chain of counting stages, giving this type of counter the name *ripple counter.*

If the clock is applied to all internal flip-flops simultaneously and internal gating determines when a flip-flop will toggle, the counter is called a *synchronous counter.* Changes in synchronous counters can occur only when the clock changes. This type of counter produces fewer "glitches" (unwanted noise pulses) than a ripple counter.

The 74160 family is also presettable. All counters have four extra inputs and also additional control input pins to control what the counter will do. A presettable counter is one that may be preloaded with data by using the preset inputs and their associated controls. We can load in a binary 5 (0101), for example, and the counter will start counting from this point. The next clock pulse into the counter will produce a 6, then a 7, and so on. If the counter is not preloaded once again, it will count to 10 (in a decade counter) or to 16 (in a binary counter) and then count the next time through in the normal fashion. Thus, presettable counters may divide by any number. A string of presettable counters can be made to divide by 1293 if this is the particular division desired.

The decade counters can be identified by their even numbers, 74160 and 74162; the binary counters, by their odd numbers, 74161 and 74163. The difference between the 74160/74161 and the 74162/74163 lies in the way in which the reset line functions. In the 74160/74161 pair, the function is asynchronous; in the other pair, it is synchronous. When the 74160/74161 counter is reset, the reset is immediate. When the 74162/74163 is reset, the reset occurs in sync with the clock.

To find out how these counters work, we will again pick the middle pair, the 74161 and the 74162. In Fig. 7-9(b), the 74161 appears on the breadboard. This is a binary counter. We will initially ignore the preset inputs and their controls. The LEDs monitor the outputs of the 74161 counter stages. The cross-coupled NAND gate generates the clock pulses for counting, allowing the clock to go slow enough for you to see what happens.

The 74161 should count in exactly the same fashion as the 7473. On the count of 16, the entire counter should reset to zero and begin the counting process all over again.

Hang another LED on the ripple carry output (pin 15). Observe the action of this LED during the fifteenth count.

Up to this point, we have been reading almost everything "upside down." If inverters are used ahead of the LEDs, the outputs can be read out when the LEDs turn on instead of off—an arrangement accomplished with a circuit called the LED buffer. The arrangement has another advantage in that the counter circuit will see the identical load each time it goes high. Since the current flow in each LED differs somewhat, the load on each counting stage is different. The input to TTL circuits, which is almost constant, is usually called "one TTL load." One TTL load (or standard load) amounts to about 1.4 milliamperes of current.

Figure 7-10 shows the circuit for the LED buffer. A low into the inverter stages will produce a high out that will turn the LED off. A high input to the inverter will cause the output to go low, thereby turning the LED on.

No pin numbers are shown on the inverter sections of Fig. 7-10. These inverter sections may be formed from 7404, 7405, 7406, 7416 or 8T90 inverters—the 7406, 7416, and 8T90 being recommended. Since the pin-outs for the 8T90 differ from the pin-outs for the others, pin designations were deliberately omitted. Use a data manual to determine the pin connections for the inverter you use to make this LED buffer.

In Fig. 7-11, the LED buffer is added to the 74161 counter circuit, and the experiment is repeated. Now when we reset the 74161 as shown in Fig.

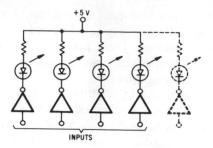

Fig. 7-10 The LED buffer

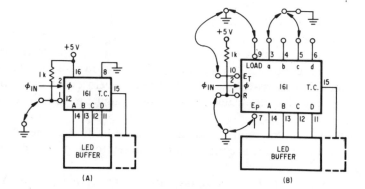

Fig. 7-11 Testing the 74161

7-11(a), all LEDs will be out and will come on as we count the pulses into the 74161.

Now let's investigate the presets, as shown in Fig. 7-11(b). At the moment, since they are all floating, they will assume a TTL high level. Use a jumper wire to ground the load terminal. Now clock the 74161; nothing should happen. Taking the load terminal low disables the counter. Now ground the Preset Enable (pin 7) as well as pin 9. Again clock the 74161. This time the highs on all the preset inputs should load into the counter. Remove the low on the Preset Enable pin *and* the low on the Load pin.

Again clock the 74161. The counter should continue counting from the decimal 15 that was loaded, go to zero, and then start counting up again. Repeat the process with preset inputs b and d grounded. This will load in a binary 0101 or decimal 5, and the counter will now start counting at "5" and repeat the pattern. You should now try other combinations of presets telling the 74161 where you want it to start counting, and you should be able to predict what the count sequence will be.

Next take pin 10, the enable toggle input, low. With pin 10 low, you should not be able to count since this pin determines whether the counter will advance or not. The Preset Enable pin determines whether the presets will be active or not. The Load pin tells the counter when to load the presets into the counter stages.

The preset enable and the count enable inputs are used in cascading counter stages. Note that four inputs control what the 74161 does. These are the rest (pin 1), the preset enable (pin 7), the enable toggle (pin 10), and the load (pin 9).

The combination of controls offered by the counter family allows some interesting things to be done. One such is use of the chip not as a counter but as a storage register. Although multiplexers will be covered in a separate chapter, we will introduce the 74150 here so that we can do an experiment with the 74161. This experiment will just fit on a solderless breadboard.

Figure 7-12(a) illustrates the circuit. The inputs shown come from the 16 switches of a calculator keyboard. Use of a calculator keyboard is not necessary, however, since all that the 16 switches do is to close a 74150 input to ground. Jumper wires can be substituted for this experiment [see Fig. 7-12(b)]. The clock circuit is the 7404 clock circuit from our earlier work. With a 0.01-μF capacitor, the circuit runs at around 200 kHz. The clock operates the first 74161, which functions as a counter. The outputs of the counter cause the 74150 to scan its inputs to see if any of them have been taken low. The outputs of the counter also feed the inputs of a second 74161, which operates not as a counter but as a storage register. Note that the

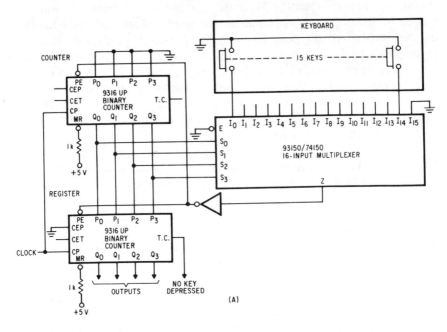

Fig. 7-12 A keyboard encoder using the 74161 both as counter and storage register (*Courtesy,* Fairchild)

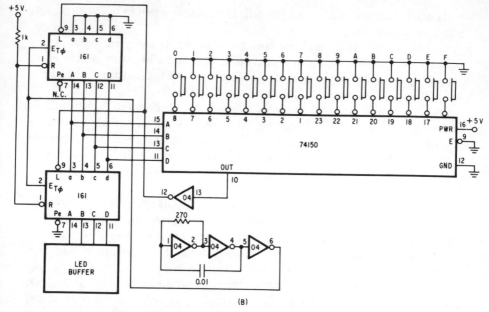

(B)

Fig. 7-12 A keyboard encoder using the 74161 (cont'd)

preset of the second 74161 has been continuously enabled by grounding the Preset Enable pin. When the 74150 detects one of its inputs low, it outputs a pulse on its output pin. This pulse in inverted and applied to the load pins on the 74161s. At this point in time, the presets have data on them that corresponds to the switch that was closed. The second 74161 loads this data into the counter stages. Four LEDs connected to the 74161 output lines will reveal the binary data that corresponds to the switch that was closed. This data is latched. The LEDs hold the data pattern until another switch is closed to produce a different data pattern, and then this new data pattern is displayed by the monitor LEDs.

What we have here is a *hexadecimal keyboard encoder*. "Hexadecimal" refers to another number system like the binary and the decimal except that it is based on 16 digits for its characters instead of just 2 or 10. We will not belabor the hexadecimal system at this time. Since it acts as a shorthand notation for the binary number system, we will read out the data from our keyboard encoder in binary for now.

The 74162 counter goes onto the breadboard for our next investigation. Connect it as shown in Fig. 7-13. Clock the 74162 through its count sequence, and observe that it is a decade counter. Now check the clear function so that you can see the difference between the members of the counter family. When the reset line on the 74162 is low, nothing will happen until you also clock the clock input. On the next clock pulse, the counter will reset, and upon releasing the low on the reset pin, the 74162 will count

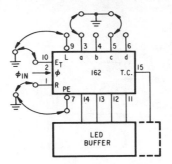

Fig. 7-13 Testing the 74162

again. Since the presets and the other control pins for the 74162 function in the same fashion as the inputs to the 74161, that portion of the 74161 experiment may be repeated for the 74162.

The 74160 is a decade counter with asynchronous clear. The 74163 is a binary counter with synchronous clear. These two members of the 74160 family will not be breadboarded since their operation may now be inferred from what you have already learned.

The 74190 Family

This group of counters is presettable, synchronous, and counts up and down. It has four members: the 74190, 74191, 74192, and 74193. The counters whose last digit is an even number are decade counters; odd last digits indicate binary counters.

Figure 7-14 gives experimental data for this family. All members have a load pin and four preset input pins. The difference between the pairs 74190/74191 and 74192/74193 lies in the method of clocking and the nature of two of the outputs.

The 74190/74191 pair has no clear input; it does have an up/down mode control, a single clock input pin, and two extra outputs (in addition to the A, B, C, and D outputs) called *ripple clock* and *maximum/minimum outputs*.

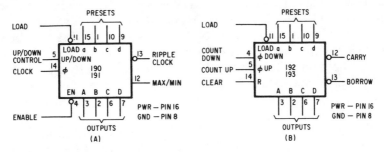

Fig. 7-14 The 74190 counter family

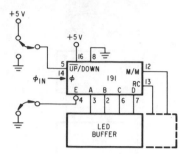

Fig. 7-15 Testing the 74191

The 74192/74193 pair has a clear input, two separate clock inputs called *up* and *down,* and two additional outputs called *carry* and *borrow.*

For our experiments, we will again pick the middle pair, the 74191 and the 74192. The 74191 goes onto the breadboard first. (See Fig. 7-15.) We will initially ignore the presets and the load control. The enable line must be jumpered to ground to enable the chip for counting. The up/down mode control pin can be controlled with a jumper as well. The clock input may be the very slow cross-coupled NAND gate, or it may be the variable clock operated slowly enough for you to observe what is happening.

The LED buffer may now be expanded to include six stages, or two additional LED indicators may be added as LED testers. The LED buffers will produce "right-side-up" read-outs; the individual LED testers, inverted read-outs. Connected as shown in Fig. 7-15, the 74191 will count up or down under the control of the input on the up/down pin.

The presets should be investigated next. These can be controlled with jumpers to either ground or + 5. The chip may be made to "auto-load" by connecting the ripple carry (RC) output back to the load pin. When the RC output goes low, it will load the 74191 with the presets. This experiment is illustrated by Fig. 7-16.

Figure 7-17 puts the 74192 (or 74193) on the breadboard. The reset line on these chips is like the reset line on the 7490s. High is reset and low is count. In addition, the line not being clocked with the up or down clock must be held high, a state conveniently accomplished by using two cross-coupled NAND gates. The switches shown in Fig. 7-17 are SPDT switches, one of whose contacts is normally closed. Although these switches are convenient, jumper wires provide the same function. Connected as shown in Fig. 7-17, both up and down clock inputs are held high by the SRFFs until either of them is pulsed low. Figure 7-17 also indicates part of the next 74192 to show how the carry and borrow outputs of the chip can be connected by utilizing additional 74192s.

The presets to this chip are handled in a fashion similar to that for the 74191. See if you can determine where the load pin should be connected experimentally to produce auto-load.

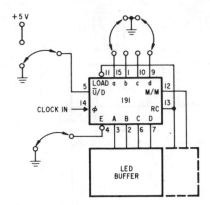

Fig. 7-16 Testing the 74191 presets

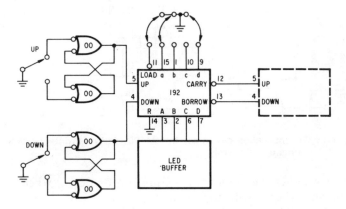

Fig. 7-17 Testing the 74192 up/down counter

Figure 7-18 shows an application circuit for the chips—an interval timer that will count to the nearest tenth of a second and will time either up or down. Although control switches are shown controlling the interval timer, in actual practice these switches would be bounceless switches. The output of the interval timer is shown going to decoders and seven-segment read-outs. Analysis of this portion of the circuitry will have to be deferred until later in the text. For the time being, the LED buffers may be used for the read-outs.

A voltage of 6.3 VAC (Volts Alternating Current) or 12.6 VAC is provided from the ac power line via a step-down filament transformer. The current into the gate is limited by the 3.3-k resistor in series with the ac. The 7413 Schmidt trigger section squares up the input ac and produces 60 PPS output. The 190/91 counter is connected to divide by 6, thereby producing 10 PPS output as the clock for the interval timer. The controls will now

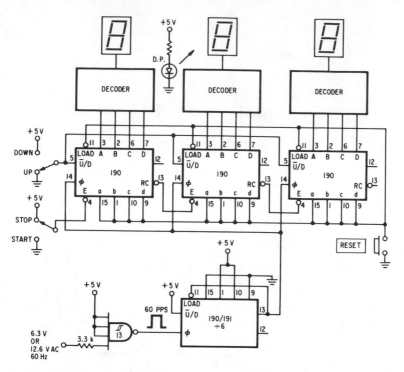

Fig. 7-18 An application circuit for the 74190 family (an interval timer that counts up or down)

determine whether the counter will count up or down, when it will start counting, and when it will stop counting.

WARNING: *A step-down transformer has 120 VAC on one side, and this voltage must be considered lethal. Use extreme caution any time you work near the power line.* Always unplug the equipment before sticking your hands into it. It is always a good idea to have a companion whenever you work around the incoming power. If you must work on a circuit with the line cord plugged in, then use only one hand and place the other in your rear pocket (the gals can place the other hand on their hip). This safety rule won't prevent a shock, but it can prevent death by electrocution.

The LED in the circuit diagram is the "decimal point." For the inputs to be on the left and the outputs on the right, the decimal point appears to be in the wrong place. However, it is correct as shown. Most circuits are drawn in such a way that if they are laid out on a PC board, the decimal points and the displays will appear in the correct places. For them to appear to be in the correct places in a schematic, it would be necessary to draw the circuitry so that the input signal enters from the right and propagates toward the left, just the opposite of what is found in conventional electronic circuitry. Therefore, some mental gymnastics are required on your part.

The 74196/8200 Family

The 7400 family of TTL chips is by far the most well-known. It is not, however, the only TTL family. Motorola had TTL 1, TTL 2, and TTL 3 families. An 8000 and 9000 series family of TTL chips is still available, and other TTL families were marketed in the past.

The 74196 and 74197 counters originated in the 8200 family. They proved to be so popular that they were assigned 7400 numbers. The 8200 family consists of decade counters (the 8280, 8290, and 8292), binary counters (the 8281, 8291, and 8293), a divide-by-12 counter (the 8288), and an up/down counter (the 8284/8285).

The 74196/97 are 30 MHz counters; the 8280/8281 are 20 MHz counters; the 8290/91 are 40 MHz counters; the 8292/8293 are low-power versions of the 8280/8281 and will count to 5 MHz. The pin-outs of all these counters are identical. For all practical purposes, the 74196/197 are the same as the 8280/8281, although the data manual does give a different counting speed for the two pairs.

These counters are presettable counters/storage latches. Because they have preset inputs, you can use them as presettable counters or as storage registers. Figure 7-19 shows the breadboard set-up for the 74196/97 (or the 8280/81) with the pin-outs for the entire family.

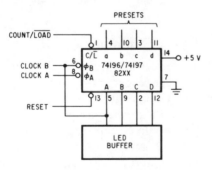

Fig. 7-19 Breadboard setup for the 74196/97 or the 8280/81 with pin-outs for the entire family of counters

This family is a ripple-counter family. The count takes place on the falling edge of the clock pulse. The counters, like those of the 7490 family, are broken into sections, with a divide-by-2 stage and a divide-by-5 or divide-by-8 stage. All members of the family have an active low reset pin and a count/load pin. To use them as counters, the two clock inputs must be connected along with power and ground, which is conventional, and to pins 14 and 7. The presets, the load pin, and the reset pin need not be connected to count. For the output of the decade counters to be in BCD format, clock input B must be connected to output A, and the clock input must be applied to clock A input. To produce biquinary division with a square-wave output, use clock B for the input clock and connect clock A input to the D output.

The presets function in a similar fashion to the presets already studied. A low on the count/load input pin will inhibit counting and preset the counter stages to agree with the preset inputs.

Figure 7-20 gives the pin-outs for the 8284/85 and the 8288. These chips will not be experimented with here, but if they should happen to be available, the pin-out diagrams and your experience with other counter members of this family will allow you to breadboard them and investigate their performance.

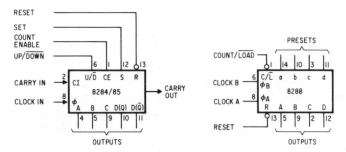

Fig. 7-20 Pin-outs for the 8284/85 and 8288

Summary

This has been a long chapter and a meaty one. Counter chips are a lot of fun to play with, but younger readers may have a difficult time appreciating how much these tiny ICs can actually accomplish. Perhaps a little history will help put things in the proper perspective. Not too very long ago a vacuum tube counter that would count to 100,000 was about half the size of a teacher's desk! Somewhat later a transistor counter was constructed that would count to 1,000,000, and the size of the package was reduced to that of a large tool box. A few more years passed, and an IC counter was constructed that would count to 30,000,000 in a package the size of a shoe box. Today, counters are available that will count to 10,000,000,000 in packages not much larger than a pocket radio.

An old catalog lists the 7400 at $58.00 each. Today, they are widely available for around a quarter. If you have to pay $1.00 for a counter chip, don't forget to appreciate what has happened in electronics over the last few years. It is about the only field where prices have strikingly decreased.

Picking a counter chip for a circuit is not a big problem. Even if presets are needed, a wide choice is available.

Chapter 8

Basic Digital Storage Devices

The world of digital electronics operates on pulses. These pulses go from low to high and back to low again, or they go from high to low and back high again. The pulses are usually very short, lasting for only a few milliseconds, a few microseconds, or even a few nanoseconds.

There are many times when the last condition of the pulse, whether high or low, must be "saved" or stored. The stored high or low can then be further processed by the electronic circuitry. To store the highs and lows, two basic types of ICs are used, the *flip-flop* and the *shift register*.

The D Flip-Flop

Figure 8-1 again returns the 7474 to the breadboard. The "D" in the D flip-flop stands for "delay." A high or a low on the D input to the flip-flop will appear one positive-going clock transition later on the Q output. It will be delayed until the clock changes, and then the state of the D input will be reflected on the Q output.

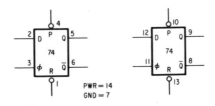

Fig. 8-1 The 7474 dual D flip-flop

As mentioned earlier, we can redefine the D input and call it a *data input*. The data, a high or a low, will then be what appears at the Q output after the clock transition. The easiest way to remember how the flip-flop

87

works is to say that whatever is on the D input will be transferred to the Q output when the flip-flop is clocked. The complement of the data will appear on the $\overline{Q}$ output.

Nibbles and Bytes

In working with microcomputers, binary digits are grouped together into units. The word *nibble* has been coined to describe four binary digits (highs or lows) treated as a group. Thus 0110 is a nibble with a value equivalent to decimal 6. Likewise, the word *byte* is used to describe eight binary digits treated as a group. Thus, 01010111 is a byte equivalent to decimal 87. Since the human eye and brain do not span eight digits easily, the byte is usually written as two nibbles: 0101 0111. The value of the byte is then represented with two digits: 0101 0111 = 57.

Now note the discrepancy. The decimal value (the number system that we are most familiar with) gives the value of 0101 0111 as 87, whereas using two nibbles to make the eight binary digits easier to read produces 57. It is obvious that the 57 is *not* a decimal value. It reflects yet another numbering system called *hexadecimal*. Hexadecimal is simply a shorthand way of writing the values of nibbles to make it easier to recall the binary code that the digits represent.

We can also break down the binary code of 01010111 into groups of three. It would then appear as 01 010 111 and be represented as 127 to simplify recalling the sequence of binary digits. This system of treating the binary digits is called the *octal numbering system*. Both hexadecimal and octal are used in working with computers. Both numbering systems are simply shorthand methods of recalling the binary code that generated the hex or octal number.

In Fig. 8-2, we are going to use the 7474 to generate a nibble. We want the binary code of 0110 to be produced. This is a decimal 6 (4 + 2). Set up the circuit of Fig. 8-2 on the breadboard. The LED buffer should be connected to the Q outputs of the 7474. If you haven't yet constructed a LED buffer, then connect the LED testers to the $\overline{Q}$ outputs so that the code generated will be *true* (right side up).

Bit $\emptyset$

In digital electronics, the numbering of data bits starts with bit zero. *Bit* is a contraction of the words *B*inary dig*it*. You will occasionally find the numbering starting with 1, but most of the time it will start with $\emptyset$. We will use the numbering scheme starting with $\emptyset$ since it seems to be the most accepted "standard," used by most authors. It really is of little consequence whether you start with $\emptyset$ or 1 so long as you observe where the numbering actually starts and don't allow yourself to get confused.

In Fig. 8-2, the D inputs of the first and fourth FF are jumpered to ground to place these bits at a logic low. The second and third bits may be

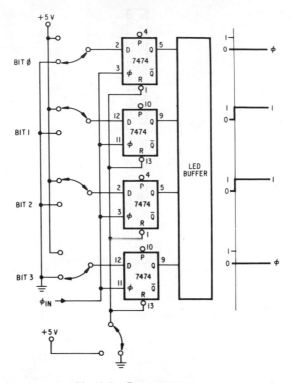

Fig. 8-2   Data storage

jumpered to plus, or they may be left floating. Left floating, they will assume a logic high. The flip-flops are now clocked. Note that they are all connected in parallel on the clock inputs and on the reset inputs. When the clock pulse rises, the data on the D inputs will appear at the Q outputs. This data will remain on the Q outputs until you change the data on the D inputs and again clock the flip-flops. The group of four flip-flops is storing a nibble.

To generate and store a byte requires that the circuit of Fig. 8-2 be duplicated. We have done this in Fig. 8-3. Now we can generate a byte. Let's produce 01010111, or 0101 0111, or 57 hexadecimal. Set up the circuit so that the least significant digit is furthest away from you and the most significant digit is closest to you. After the circuit is operational, and the appropriate D inputs have been jumpered to ground and the group of FFs clocked, the entire breadboard can be rotated 90 degrees clockwise, and the read-out of the digits will then be the same as the binary code listed above.

The 74175

The former experiment took four 7474 chips to get eight flip-flops and their D inputs. The 74175 is a group of four D flip-flops in one package, as

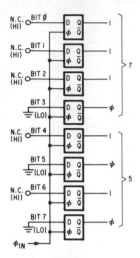

Fig. 8-3   Storing a *byte* of data

shown in Fig. 8-4(a). Thus only two 74175s will be needed to accomplish the same thing. Figure 8-4(b) shows the same experiment using two 74175s.

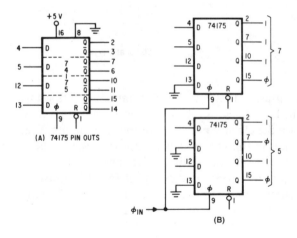

Fig. 8-4   The 74175 Quad D storage register

Newer chips are available that have eight flip-flops in one IC package. The main reason that one of these is not being used here has to do with cost. The two 74175s or even the four 7474s cost less than the single package of eight flip-flops. The single chips will require less PC board real estate in commercial production, but for experimental purposes we are better off using more ICs if doing so costs less.

Shift Registers

The data generated and stored in the experiments in this chapter so far has been handled in parallel, that is, all eight bits of data have been generated and stored at one time. There are times when a single bit at a time is generated and stored. Eight bits can then be stored in this fashion and treated as a byte. The device used for this purpose is called a *shift register*. Since either the 7474 or the 74175 may be used for the next experiment, Fig. 8-5 shows four D flip-flop sections without any pin connections.

To see the shift register in operation, we first must get a pulse into the first D FF. Start by clearing all FFs by grounding the rest line momentarily. Then use the preset input to set the first Q output high. Now clock the FF chain. The high pulse will be observed traveling down the FF chain as the chain is clocked.

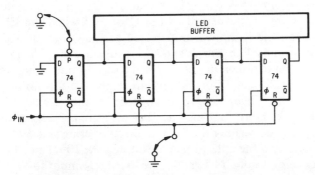

Fig. 8-5 The 7474 in a shift-register configuration

To get the initial high into the 74175s that do not have a preset input, we must momentarily lift the first D input from ground, clock the FF to get the first high in, then ground the first D input again as we send the high down the shift register.

To send a low down the chain, we must first get all Q outputs high. This can be accomplished with the preset inputs for the 7474s. For the 74175s, lift the first D input and clock the chain four times to set all Qs high. Then enter the low bit on the first D input and reconnect the first D input high again. Now as the flip-flop chain is clocked, the low will be observed traveling down the FF chain. And, of course, the $\overline{Q}$ outputs of each D FF can be connected to the following D inputs to make a shift register that will propagate the low down the chain. We'll leave this experiment for you to draw out, then set up on the breadboard and perform.

D FFs are not normally used for shift registers. There are a wide variety of 7400 shift register ICs available for this purpose. Some of them have only four flip-flop stages, but some have five or eight FFs built into the chip. The data enters the shift registers in serial form and is extracted in parallel. This type of shift register is called a *serial-to-parallel converter*. The reverse type shift register, for conversion of parallel to serial, is also available in the 7400 series.

A UART, or Universal Asynchronous Receiver-Transmitter, is a special chip that has both serial-to-parallel shift registers and parallel-to-serial shift registers packaged in the same IC. It also has many additional "control" pins that allow you to specify the number of bits to be processed, the format that will be used, and even the clock rate that will accomplish the parallel-to-serial and serial-to-parallel conversion. Of course, the conversions are abbreviated: S/P means serial-to-parallel and P/S means parallel-to-serial conversion.

The JK Flip-Flops

Earlier we ignored the J and K inputs to the *JK* Flip-Flops. We return the 7473 to the breadboard now to investigate the J and K inputs. Whatever is on the J input is transferred to the Q output when the FF is clocked. Whatever is on the K input is transferred to the $\overline{Q}$ output when the FF is clocked. It sounds like the J and K inputs are double D inputs. Try connecting them together and using them as a D input. When you clock the FF, you will find that if both inputs are high, the flip-flop operates normally, that is, it changes state for each clock. If they are both low, the flip-flop won't work. Thus, they may not be connected and used for a D input.

In Figure 8-6, we have the 7473 set up to operate as a shift register. The J input is used and the data propagates down the FF chain. The K input is tied high or left floating. In Fig. 8-7, we use the K input to transfer the data down the chain using the $\overline{Q}$ output. This time the J input is kept high.

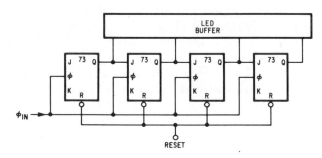

Fig. 8-6   The 7473 in a shift-register configuration

In Fig. 8-8, we show an application of the JK flip-flop and the D flip-flops to take in two nibbles and convert them to a byte of data. The input data lines connect to both 74175s in parallel. The JK flip-flop will enter the data into one of the 74175s when the Q output rises on the JK flip-flop and enter the data on the input lines in the other 74175 when the $\overline{Q}$ output rises.

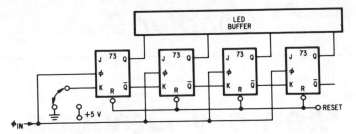

Fig. 8-7 Using K and Q

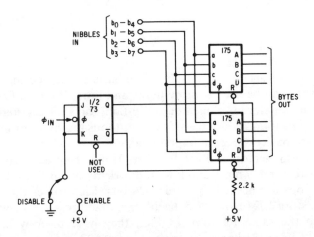

Fig. 8-8 Converting nibbles to bytes

The two nibbles stored in the Q outputs of the 74175s may then be extracted as a group of eight binary digits, that is, as a byte of data.

Summary

A great many different flip-flops and shift registers are available in the 7400 family. All these devices may be used to store the highs and lows of digital data. A nibble is a binary group of digits with four highs and lows. A byte is a group of eight binary digits treated as a whole.

Chapter 9

Memory Devices

In Chap. 8, we learned that flip-flops are basic storage devices or memory devices. They can be used to store highs and lows. If only a few nibbles or bytes are to be stored in memory, flip-flops or shift registers provide a satisfactory method of storage. If several thousand bytes of data must be stored, these devices become impractical.

The basic memory cell is a flip-flop. It may be used to store a high or a low. To store many highs and lows requires many flip-flops. For example, 1024 flip-flops are needed to store 1024 bits. Eight times this number are needed to store 1024 bytes of data. IC manufacturers can pack many flip-flops onto the same IC chip. In fact, they have recently increased the number of FFs packed onto the same chip to 8192 and in all probability will pack even more within the same IC chip in the future. The chips with all these FFs on them are the memory chips. The type of memory chip that we have been describing is called a *static memory chip*. The data stored in the FFs will remain as long as power is continuously applied to the chip. Once power is removed, the stored data is lost. This type of memory is called *volatile*.

Another type of memory chip is called the *dynamic memory chip*. In this chip, the data is not stored in a flip-flop but is stored as a charge. Because the charge can "leak off," the dynamic memory cell must be periodically "refreshed" to restore the charge lost by the electrons leaking away. This type of memory cell requires fewer transistors and is thus less expensive. The static memory cell requires no refresh circuitry.

ROMs and RAMs

If we take a pair of 74175s and wire the D inputs so that the pair will always produce the same code pattern, we have what we call a Read Only Memory, abbreviated as ROM. Since the D inputs are permanently wired in one configuration, the 74175s will always produce the same binary code pattern. If power is removed, and then reapplied, the same code pattern will be retained only if the data is clocked in, since there would be no guarantee that the flip-flops would come up in the "correct" state when power is applied. A true ROM would not require that the data be clocked back in. The data is permanently fixed in the true ROM and is immediately available as soon as power is applied. Of course, we usually do not use bunches of 74175s for creating a Read Only Memory. There are special ICs available for this purpose.

If we operate the 74175s as shown in Fig. 8-8 of the preceding chapter, the data stored by the 74175s will depend on what is on their D inputs when we clock them. Since we can change this data at will, the circuit of Fig. 8-8 illustrates a Read/Write memory. The Read/Write memory is volatile, however, since removing the power and then reapplying it will not reproduce the same code pattern out of the 74175s that was there before the power was removed.

If a group of flip-flops is packaged on an IC chip with address and decoding circuits so that any single FF or memory cell can be selected by changing the address and decoding data, the memory chip is said to have *random access*. This means that any memory cell in the entire array can be individually selected at any given time. ROM and RAM both possess this random-access capability, but somehow the literature got filled up with use of the word *RAM,* which stands for Random Access Memory, to describe or identify the Read/Write memory and use of the word *ROM,* or Read Only Memory, to describe or identify the memory that has fixed data patterns. We won't fight the problem, but it does tend to confuse the newcomer somewhat. Both RAM and ROM are random-access devices, and any single memory cell in the array may be individually selected. Some ROMs are erasable. They are usually designated as EROMs or by some other terminology to differentiate them from ROMs that have their pattern fixed during manufacture and cannot be changed by the user. RAM, or Read/Write memory, is memory that is volatile but may be written into and read from at will. ROM is nonvolatile but is written into once and then only read from thereafter.

The 7488

To investigate the ROM, we will use one of the least expensive PROMs available, the 7488 or the 8223. A PROM is a Programmable Read Only Memory. The 7488 is a PROM that can be programmed "in the field," which means that *you* can program it. This PROM is a fusible-link prom, which means that to program it, the internal fuse links must be blasted apart. It is normally supplied with all outputs at a logic zero. Whenever a logic one is needed, the internal fuse link must be blasted, or burned open. This is the origin of the term "burning a ROM."

Before the PROM is burned, the pattern to be entered must be determined. Although we have yet to discuss the hexadecimal number

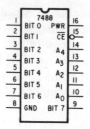

(A) 7488/8223 PIN OUTS

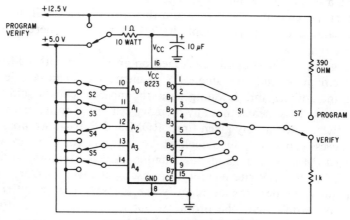

(B) SIGNETICS MANUAL PROGRAMMER CIRCUIT

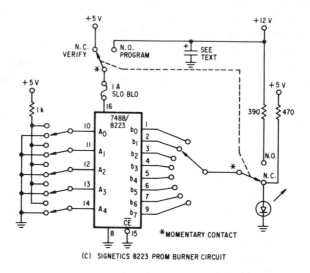

(C) SIGNETICS 8223 PROM BURNER CIRCUIT

Fig. 9-1   The 7488/8223

system in detail, we will be using it eventually and will need a decoder to convert the binary code pattern to seven-segment code. There are ICs available for this purpose, but we can make our own hexadecimal seven-segment decoder from the 7488.

The starting point for burning the PROM is to create a data table that is to be entered into the ROM. Figure 9-1(a) gives the pin-outs for the 7488/8223, and Fig. 9-1(b) shows the Signetics Manual Programmer circuit for the 8223. Fig. 9-1(c) is the author's version of the 8223 Manual Programmer. Figure 9-2 shows one method of organizing the data to be entered into the ROM. The 7488 is organized as 8 bits X 32 words. This means that there are a total of 256 bits available for the ROM, and these bits are arranged in a matrix inside the 7488 that is 8 wide and 32 deep, as reflected in the table organization of Fig. 9-2.

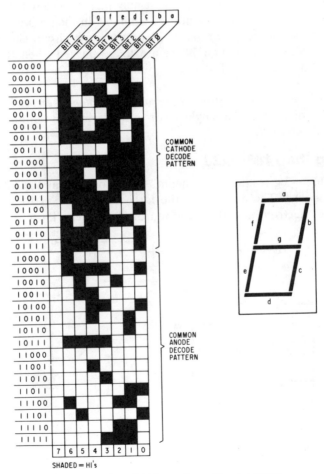

Fig. 9-2 The PROM matrix and burn pattern

Each horizontal row in Fig. 9-2 is called a *word*. The word length is eight bits. To address (select) a row (word), a binary code pattern is entered on the address pins. A binary code pattern of 00000 will select the top row, whereas a code pattern of 11111 will select the bottom row. As this address pattern is changed, each row may be addressed.

Each column in Fig. 9-2 represents one of the eight bits in the word (row). In burning (programming) the 7488, the row address must be selected and the bit must be selected. In the Signetics 8223 PROM burner circuit of Fig. 9-1(b), this is accomplished with five toggle switches for the addresses and an eight-position rotary switch for the bits.

Although it is not an easy task, and extreme care must be taken, we can program the 7488 on the breadboard. The switches are all replaced with jumper wires. The + 12 V required for the burning operation can come from a 12-V battery out of the family car, it may be a 12-V power supply from Chap. 13, or it may be a 12-V supply available in the school lab. If many 7488s are to be burned, then the 8223 programmer circuit should be constructed. If only one or two 7488s are to be burned, then these can be programmed on the breadboard using jumper wires to control the rows and columns.

One final word of caution. *An error is a permanent error.* The 7488 will have to be discarded if a single error is made in programming. Proceed slowly and carefully.

Burning the 7488/8223

Figure 9-3 gives the experimental set-up for burning the 7488/8223 on the breadboard. The purpose of the large capacitance is to store a large quantity of electrons near the breadboard. The battery (if used) may then be

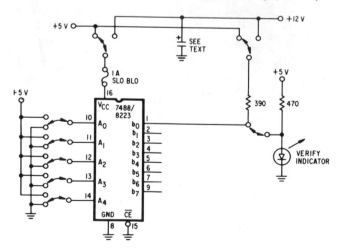

Fig. 9-3   Burning the 7488/8223 on the breadboard

placed on the floor some distance away from the breadboard. The capacitor is charged up from the battery, and the energy stored in the capacitor is then "dumped" into the fuse link.

Start with the first row and with bit $\emptyset$. Figure 9-3 is set up for this starting point. The fuse in the circuit will prevent total destruction if the students get wires in the wrong places. Now, double-check all connections. Remove + 5 V from pin 16. Connect the bit to be burned through a 390-ohm resistor (not critical; anything close will work) to the + 12 V source. Now momentarily touch pin 16 to the + 12 V source. Reconnect + 5 V to pin 16. Connect the LED tester to this bit to verify that you blew the internal fuse link. This bit should now be high; if it is not high, you did not blow the link, and the burn process will have to be repeated. Some 7488 fuse links can be quite stubborn. Keep trying until you burn the link open. Keep taking the chip's temperature with your finger tip. When the chip gets hot, you will have to stop trying to burn it and let it cool off. The internal resistance of the fuse link increases with temperature, making it more difficult to blow. Continued attempts to blow the increasing resistance will simply make the chip temperature continue to rise, compounding the problem.

Occasionally, you will encounter a fuse link that defies blowing. During the manufacturing process, the nichrome may have been deposited a little too thickly. When this happens, you are justified in increasing the burn voltage until the internal link blows. If it won't blow, nothing will be lost since the chip will have to be discarded anyway. Increasing the burn voltage causes the chip temperature to increase rapidly; therefore, keep the earlier comments concerning temperature and resistance in mind if it becomes necessary to do so.

Once the first bit is successfully burned, you proceed to the next bit. The process is repeated for each bit in the matrix that must be changed to a high. When the first row (word) has been completed, the row address is changed to the next row by changing the jumpers as required, and the bits in this row are then burned.

Sixteen rows need to be burned for the seven-segment decoder. If you bungle the first 16 rows with an error, you can try again in the second half of the chip. If address line A4 is high, you are in the second half of the chip. Try again.

Seven-segment displays come in two varieties. One type needs active highs out of the bits, whereas the second type needs active lows. One 7488 will provide both types of seven-segment decoders *if* you were successful in burning the first 16 rows correctly. The burn pattern for the second 16 rows (the other type of seven-segment read-out) will be just the opposite of the first 16 rows. If you were not successful in burning the first 16 rows correctly, you can try again in the bottom 16 rows. The bottom 16 rows of the PROM matrix are accessed by making address line A_4 high so that you get two chances to do the job correctly. You can have both types of decoders

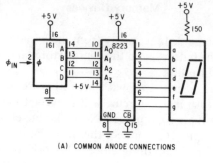

(A) COMMON ANODE CONNECTIONS

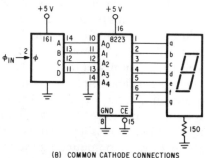

(B) COMMON CATHODE CONNECTIONS

Fig. 9-4 Connecting the homebrew decoder

in a single 7488/8223 only if you are successful in programming all 256 bits of the matrix correctly. Once the seven-segment decoder is programmed, it may be tested with a seven-segment readout. Figure 9-4(a) gives the connection for the common-anode type of seven-segment readout, and Fig. 9-4(b) for the common-cathode type.

A large variety of ROM chips are available. Those programmed at the time of manufacture are called *mask programmable*. A factory will program a ROM for you, but the mask set-up charge will amount to several thousand dollars. If you need 100,000 chips, this charge can be spread out and be economically feasible; but if only one or two are needed . . . .

Another type of available ROM is the EROM, a Read Only Memory that is field programmable and also erasable. It is usually supplied with all its bits at a logic high. When a bit is programmed, it can be changed to a low as required. If an error is made, these ROMs may be erased with a special ultra-violet light source and reprogrammed. We did not use them here because they are more expensive and require special programming equipment; they are therefore more difficult to program on an experimental basis. Yet another type of ROM is the EPROM, an Erasable Programmable Read Only Memory.

Read/Write Memory

To study the Read/Write Memory, we will use two chips, the 7489 (8225) and the 21L02. The 7489 is a TTL RAM organized as 16 words $\times$ 4 bits. We will connect two of them to make a small memory of 16 × 8. The same pattern that was entered into the 7488 will also be entered, and for the same purpose—to provide the seven-segment decode function. After going to all the effort of entering the bit pattern into the RAM and then verifying that it works properly, we will deliberately turn off the power to the RAM and then turn it back on again. This will quickly demonstrate what is meant by volatility, and with a lasting impression!

The 7489 (8225)

Figure 9-5(a) shows the pin-outs for the 7489; Figure 9-5(b) shows the test circuit. Two 7489s are required to get a matrix eight bits wide. Data out of the 7489 is inverted. This means that the inputs are inverted in the chip and exit as the inverse of the state in which they were entered. We could hang inverters on all output lines, but we can also use the "opposite" bit pattern for the seven-segment decoder and accomplish the same thing.

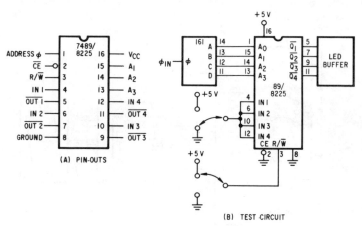

Fig. 9-5 The 7489/8225 RAM

We will use Fig. 9-6 to introduce some concepts from the computer field. A computer has three major groups of signals. One group consists of the address lines (computers frequently have 16 address lines). A second group of signals consists of data. A third group of signals consists of control signals.

When signals are treated as a group, we use the term *buss* (also spelled *bus*). The address buss consists of 16 address lines treated as a single group of signals. The data buss consists of eight signals treated as a group. The group of signals that control the operation of the computer forms the control buss. The control buss may have only one signal, or it may have several.

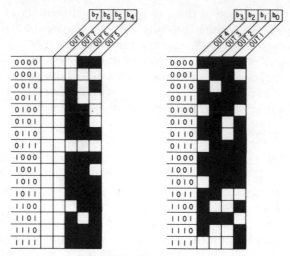

(A) THE 7489 INTERNAL MATRIX

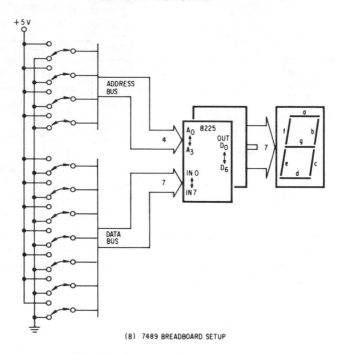

(B) 7489 BREADBOARD SETUP

Fig. 9-6   Creating a decoder in RAM

As the circuit density becomes greater, drawing in all lines for the address buss, the data buss, and the control buss may make the circuit diagram more difficult to read. To simplify things, a wider line is used to depict the many wires of a buss, although a single line is still used for a single conductor. A number is often appended to the wider line to indicate the number of wires in the buss, but not always because it is usually understood that there are 16 lines in an address buss and eight lines in a data buss.

The 8225 in Fig. 9-6(b) can be controlled with jumper wires. Four jumper wires provide the address lines. We need only four wires because the matrix involved is only 16 rows deep instead of 32 as in the 7489. Since the data buss is eight bits wide, eight jumper wires will be needed to encode the data lines. Since only seven of these are used, one jumper can be left connected in the same position throughout the experiment. The seven outputs of the two 7489s can be monitored with seven LEDs, with LED registers, or with a seven-segment readout.

The $\overline{CE}$ line must be connected to ground with a jumper ($\overline{CE}$ stands for *Chip Enable*). Many ICs used with computers have a CE pin. This pin allows the chip to be "removed" from the electronic circuit by applying an appropriate high or low to it. If the CE input is not accompanied by a small circle, it is an active high enable. Taking the pin low will disable the chip. If the small circle is present, or if the chip enable is overscored ($\overline{CE}$), then the input is an active low, and leaving the pin open (floating) or taking it high will disable the chip. Many chips used in computers have more than one CE pin. They may also have a CE and a $\overline{CE}$ pin on the same chip, or even combinations of the two. These multiple CE inputs allow the chip or groups of chips to be selected (enabled or disabled) with less gating, thus helping to keep the cost and the complexity of the circuitry down.

Since the 7489 is an open-collector chip, pull-up resistors must be added to the outputs.

Read/Write, or RAM, chips also have another input that we have not come across as yet. This is the $R/\overline{W}$ input. When this input on the 7489 is high, the chip is in the read mode, and data will be read out of the matrix. When this input is low, the chip is in the write mode, and data will be written into the memory cell that is addressed at the time the $R/\overline{W}$ goes low. On different types of RAM chips, this R/W line may be $\overline{R}/W$ or $R/\overline{W}$. This signal out of the microprocessor chip itself may also be $\overline{R}/W$ or $R/\overline{W}$. To completely control this signal among all the different memory chips may require the use of several inverters in the circuitry to get the correct polarity signal for the computer system.

In Fig. 9-6(b) we will start with address $\emptyset\emptyset\emptyset\emptyset$. Ground all address lines. Let's use the top 7489 for the left portion of the matrix and the lower 7489 for the right portion of the matrix. To encode a 0 on a seven-segment read-out, all segment lines except that for the g segment must be either high or low. Whether it is high or low will depend on the particular type of seven-segment read-out used. We will select the decode sequence for the FND 70. This is a common-cathode type of seven-segment read-out. In the common-cathode type of read-out, the segments are turned on with a high to each segment. We therefore need all segments except g high. Since the 7489 inverts, we need to place a high on the g segment line and have all other segments tied low. The R/$\overline{\text{W}}$ line is then taken low to write this data pattern into the 7489 in the first row. When we return the R/$\overline{\text{W}}$ to high to return to the read mode, the pattern displayed should be 00111111, or, if a seven-segment common-cathode display is used for the read-out, the numeral ⌀ should be displayed.

The address line is now changed to ⌀⌀⌀1 by changing the appropriate jumper; the data pattern for a 1 is set up on the data input lines (segments b and c, high; the rest, low); and the chips are again written into. The resulting output should now be 00000110, or the numeral 1 should be displayed on the seven-segment readout.

Proceed to encode the rest of the seven-segment decode pattern into the 7489 matrix. The nice thing about RAMs is that errors are easily corrected. When you have the entire array programmed with the decode pattern, the four address lines may be removed. The output of a binary counter such as the 74161 can then be connected in place of the wires used for the address jumpers. If the counter is now driven with a slow clock, the display will cycle through all the numbers of the hexadecimal number system, and you can watch your handiwork being displayed automatically. The address lines of the 7489 could have been more conveniently controlled by using the counter, but the more complex the circuit, the easier it is to lose a reader in it. But for this objection, we could even have taken the 7489 already programmed and used it and a counter to control the data lines. Thus we have used jumper wires in an attempt to keep things as simple as possible until you get things under control. If you would now like to try using the 7489 and the counter to control the address lines for encoding, the circuit is given in Fig. 9-5(b). Without a console and some of its support circuits, you may experience a shortage of breadboard area in this experiment.

When you have thoroughly tested the 7489s and are satisfied that you can now program RAMs, then disconnect the power to the console and then turn it back on. All your work will have vanished, and you will now know what is meant by a *volatile* memory. (In any practical sense, of course, once you had a RAM encoded with the data patterns desired, you would "save" the data by storing it in a cassette or floppy disc or by transferring it to a ROM. The next time you wanted to use the coding, you would not have to do all this work over again.)

When either RAM or ROM is used in the fashion investigated here, we make the memory serve as a look-up table. In this instance, the conversion of the binary data is from an address to a hexadecimal character display. We enter a binary code pattern as an address. The output of the circuit appears as a hexadecimal number on a seven-segment display.

This concept of the look-up table applies equally to any other conversion from one code to another. The same technique is applicable for converting binary to other codes, for converting Baudot (Murray) to ASCII, for converting EBCDIC to ASCII, for converting Selectric Tilt and Rotate to ASCII, or for converting whatever else needs to be translated from one data base to another.

Memory Fetch

One of the more difficult concepts for the newcomer to comprehend is the idea of memory fetch. When an address is placed on the address buss, memory returns data to the microprocessor. This is called *memory fetch*. The experiments on the 7488 and on the 7489 have already illustrated it. We placed an address on the ROM or the RAM, and the memory device immediately returned the data stored at that address.

It takes time for the address to settle into the ROM or RAM, and time for the ROM or RAM to respond to the address. This is referred to as *memory access time* and is about a microsecond or less for the majority of currently available memory chips. For the TTL memories used in this experiment, the access time is on the order of nanoseconds. Although you are not aware of this small delay in the return of data, it does exist. For all practical purposes, the memory instantly returns data requested of it with the address buss.

The 21L02

Sixteen bytes of RAM is a pretty small memory; 1024 bytes gives us a considerably larger one. Just as it took two 7489s to get a memory matrix eight bits wide, the 7489 being organized as 4 X 16, it takes eight 21L02s to get a memory matrix eight bits wide because the 21L02 has a 1024 X 1 organization.

Eight chips of 16 pins each will just fill one Superstrip. We can get eight 21L02s on the breadboard, but there will be no room left for any support chips. In the classroom, the students can pair up and have two Superstrips available. For those of you working at home, you can either invest in a second Superstrip or you can "stack" the RAM. All address lines and all $\overline{CE}$ lines of the eight 21L02s connect in parallel. Only the data lines need be separated. It is therefore possible to stack the 21L02s one on top of the other and very carefully solder each pin to the corresponding pin on the chip underneath it. Wires are then attached to the data-in and data-out pins, and these wires are plugged into the holes on the Superstrip. The space taken up by the RAM stack will then be reduced to approximately the space required for two 16-pin chips, leaving the remainder of the Superstrip available for the rest of the support circuitry. This technique works satisfactorily, but it does mess up the pins quite a bit, and, if you find a bad memory cell in a chip in the middle of the stack, it is quite a chore to change the chip. Consequently, this technique can be recommended only as an emergency alternative.

Figure 9-7(a) gives the pin-outs for the 21L02. The 21L02 was selected because it is one of the least expensive 1K memory chips available. It is also one that is commonly used in computers. Figure 9-7(b) gives a test circuit for the 21L02. This is a go-no-go test circuit. It will not reveal some chip defects, such as two address lines being shorted, but if the 21L02 fails this test, it is definitely defective and may be discarded.

Figure 9-7 uses three binary counters to generate the 1024 addresses needed for the 21L02. Any binary counters may be used, and even dual flip-flops may be substituted. Initially, the input clock to the counters can be the cross-coupled NAND gate, but this will prove to be far too slow. Variable speed clocks will be employed to test the 21L02 past the first few memory cells. The data input line to the 21L02 is controlled with a jumper, which is connected either to ground or to + 5 V. The R/$\overline{\text{W}}$ line is also controlled with a jumper: high for read, low for write.

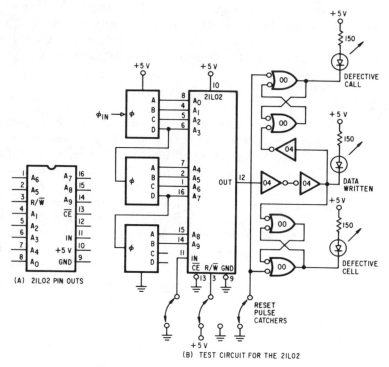

Fig. 9-7   The 21L02 and test circuits

The output of the 21L02 is shown going to an array of gates, inverters, and LEDs. These form a pulse catcher circuit. The middle LED will indicate the data out of the 21L02. It will be on if the 21L02 has been filled with all highs, and off with all lows. It will be steady if there are no defective memory cells in the 21L02. It will flash or flicker if the chip has defective cells.

The other two LEDs along with their associated circuitry are pulse catchers. After the 21L02 has been written either all high or all low, the Reset line of the pulse catchers is momentarily grounded to reset the two pulse catchers. Both LEDs connected to the cross-coupled NAND gates should go out. The 21L02 is then clocked with the counter circuits, and all 1024 addresses are scanned. If the 21L02 has no bad memory cells, the inputs to the cross-coupled NAND gates remain steady, and they are not triggered. If the 21L02 has a single defective memory cell in the array of 1024 cells, the pulse generated by this cell will trigger the latches and one of the two LEDs will turn on. One LED will light when testing for all highs being written into the 21L02, and the other will light when testing for all lows. If the chip under test has no defective cells, the two pulse-catcher LEDs will remain off during both tests. To test 1024 memory cells individuallys takes quite a long time. Use of pulse catchers to catch any defective cells, however, allows the counters to be operated at high speed (up to 1 MHz).

To test the test circuit and see how it works, you must first get everything connected. Then reset the counters to 0000000000. The clock input to the counters should initially be the single clock pulses produced by the cross-coupled NAND gate. Ground pin 11 and enter lows into the first 16 or so memory cells. To write into the chip, the $R/\overline{W}$ line must be taken low. After 16 or so cells have been written low, connect the data input line (pin 11) high and write just one cell high; then return the input line to ground and write several more cells low.

Now place the chip in the read mode by taking the $R/\overline{W}$ high. Reset the counters so that you can scan the cells already written into. Now reset the pulse catcher circuitry by taking its reset line momentarily to ground and clock the counters again. Nothing should happen to the LEDs until you get to that one high in the string of lows written into the memory cells. The one LED monitoring the data should go high, thus indicating the high cell. At the same time, one of the two pulse catcher LEDs should turn on. As you continue clocking the counters, the monitor LED will again go out, but the lit pulse catcher LED will remain lit, indicating that one of the group of memory cells being tested did not have the same data in it that all its neighbors did.

Now repeat the experiment using a high-speed clock. The 200-kHz clock used in one of the earlier experiments (7404 clock with 0.01 feedback capacitor) can be used if you haven't constructed one of the variable clocks yet. Feed the 21L02 with all highs or all lows. At 200 kHz, all memory cells will be written to in less than a second. Now use the slow clock, and enter just one cell with the opposite data. Then reset the pulse catchers, and again run the circuit at high speed. The pulse-catcher circuitry will again capture that one defective cell in the array of 1024. You don't even have to reset the counters to zero since they will scan the chip several times before you can shut things down.

Now that you have the pulse-catcher circuitry operational and can see how it functions, you can test the 21L02. Write all lows into the chip. Then reset the R/$\overline{W}$ to read, and the pulse catchers to extinguish, the pulse catcher LEDs, and turn on the clock to the counters. If the 21L02 has no defective cells, the pulse catcher LEDs will remain off. Now change the R/$\overline{W}$ back to write, change the data input line to a high, and write all cells high. Again return the chip to read, reset the pulse-catcher circuitry, and again clock the 21L02s. In less than a minute, you will have written all 1024 cells high and checked them, then written all 1024 cells low and checked them as well. If you try to test 1024 cells for both highs and lows by going slow enough to see the LEDs flicker, the classroom bell will ring before you can get them all checked.

As we mentioned, this is not an exhaustive test of the 21L02. If it failed this test, then it should not go into a computer. If it passed, it can go into the computer for additional testing.

Building Computer Memory with the 21L02

How do you build a computer memory? What is meant by 4K of memory? How can you draw eight 21L02s interconnected to form a 1024 $\times$ 8-bit memory block and not have the drawing so small that it strains the eyes to try to see where all the lines go?

Figure 9-8 shows one method that can be used. Even here it is necessary to abandon the convention that says all inputs must enter from above or from the left and all outputs must be from the bottom or from the right. The eight 21L02s have been drawn in a stack, one behind the other, to convey the idea that eight chips have been used to construct "1K" of memory. (The "024" in "1024" is dropped and the number rounded off to 1000; the three zeros are then replaced with the letter K, indicating 1000. Thus 4K of memory actually consists of 32 21L02 RAM chips, eight chips being used for each 1K of memory. The total number of *bits* is 4096 $\times$ 8: the total number of *bytes* is 4096; and the block as a whole is referred to as 4K of memory.)

The ten address lines required (A_0-A_9) are drawn as a wide line and marked with a 1$\emptyset$. The data buss is split, with data coming into the 21L02 on one buss, and data going out of it on another. A third buss is called the control buss, which, in the case of 1K of RAM, requires only two signals. All busses are indicated as proceeding to a location other than the 1K block of 21L02s.

To build each 1K of RAM memory requires eight 21L02s. If this is the only memory in the system, then the $\overline{CE}$ is grounded. If more memory is present, the $\overline{CE}$ line must be fed from a decoder. (Decoders will be discussed

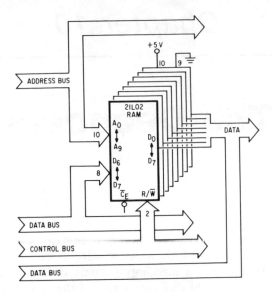

Fig. 9-8   1 K of RAM using 21L02s

in the next chapter.) Moreover, all lines to the chips other than the inputs and outputs must be paralleled.

To have put the words "data in" and "data out" in Fig. 9-8 would have been an error. "Data in" and "data out" are always referenced to the microprocessor and not the memory chips.

Summary

Any device that will store a high or a low is a memory device. Memory devices may be static or dynamic. They may be Read-Only-Memory (ROM) or Randon-Access-Memory (RAM). Since both ROM and RAM are random-access devices, a better term for RAM is *Read/Write memory.*

Memory chips are organized into different matrices. The 7488 has a 32 × 8 organization; the 7489 has a 16 × 4 organization; the 21L02 has a 1024 × 1 organization. It takes eight bits to make a byte. To make a matrix of 1024×8 memory cells requires eight 21L02 chips. The 1024 bytes of memory in the block are defined as 1K. A memory of 4K means 4096 × 8 bits of memory cells.

Chapter 10

The Decoders

A decoder is a device that accepts a binary code pattern input and outputs a different code. Figure 10-1 shows a decoder made from gates and inverters. The circuit accepts a three-input binary code and produces a low output for each of the eight different states of the code (two input types, either high or low, for three different input lines, yielding 2^3, or eight different states). This device is called a *one-of-eight (1:8) decoder* or *de-multiplexer*. We will call it a *decoder* for now and save the multiplexers and demultiplexers for the next chapter.

Fortunately for us, we don't have to build up the circuit of Fig. 10-1 from gates and inverters. The 7400 family offers many decoders. One group consists of one-of-ten (1:10) decoders. This group takes a four-bit binary code input and puts out one low for each input code combination. These would be used to drive Nixie® tubes. Another group takes the input binary

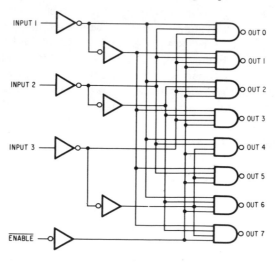

Fig. 10-1 A decoder made from gates and inverters

code and outputs the appropriate code for operating seven-segment read-outs. Other decoders accept the binary input code and output a different code, such as the Excess Three Grey code.

The 7442

Figure 10-2 illustrates the 7442. The 7442 is a 1:10 decoder. It may also be operated as a 1:8 decoder or any decoder with less than ten outputs. We will start by operating it as a 1:4 decoder. Connect it as shown in Fig. 10-2(b). The binary code input will initially be supplied by jumper wires. Note that the LED connected to the zero output lights when all inputs are low, corresponding to the input code of ∅∅∅∅. Now enter a code of ∅∅∅1 by lifting the "a" input to the 7442. The LED connected to the zero output should go out, and the LED connected to the one output should come on. As the input code is changed, each of the decoded LEDs should light.

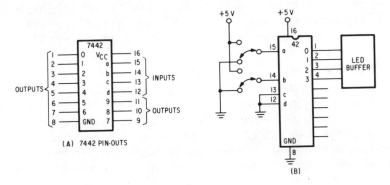

Fig. 10-2 The 7442 operated as a 1:4 decoder

In Fig. 10-3, we operate the 7442 as a 1:8 decoder. The jumper wires are replaced on the input circuitry with any of the binary or decimal counters. As the counter is clocked, the binary output code is fed to the

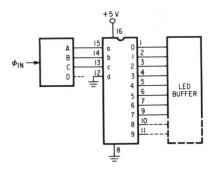

Fig. 10-3 The 7442 operated as a 1:8 decoder

7442, producing eight different outputs. We need to connect eight LEDs to the 7442 outputs. Each LED will turn off as the decoded output goes low. If discrete LEDs and current-limiting resistors are used instead of LED registers, the LEDs will light one at a time as the outputs go low. We can also place inverter sections between the 7442 outputs and the LED register inputs to cause each LED in the LED register to turn on when decoded instead of off.

By connecting the "d" input of the 7442 to the D output of the counter, the 7442 can be operated as a 1:10 decoder. Two more LED indicators must now be added to the two additional outputs to see them being decoded. The counter need not be a decimal counter. For all inputs above 1001 (decimal 9), all outputs of the 7442 remain high.

Decoders may be operated as 1:5, 1:6, 1:7, 1:9, and so forth simply by not using the additional outputs. The binary input code has to have enough states to scan the internal decoding circuits completely. To operate the 7442 as a 1:6 decoder, the basic 1:8 decoder circuit must be used with the required three binary input lines. To operate the 7442 as a 1:9 decoder, all four inputs of the 1:10 decoder circuitry must be used; the last output is simply ignored.

The 74154

Figure 10-4 combines the·pin-outs and the working drawing in one. The 74154 is a 24-pin chip that uses all four binary inputs and outputs 16 different decoded lows for each input combination. Either jumper wires or a binary counter may be used to provide the binary input code. The 74154 has two enable lines, *both* of which must be low for the decoder to function. LEDs may be attached to the 16 output lines to monitor the decoder function.

The number of LEDs needed to monitor the output lines becomes

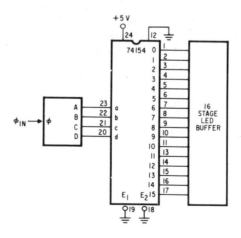

Fig. 10-4   The 74154 1:16 decoder

awkward. Figure 10-5 shows one possible and very handy alternative. Take a 16-pin socket, place eight miniature LEDs across it, and plug them into the socket connections. Using a hot glue gun, run a bead of glue along each side of the LEDs to hold them firmly in position. Since the glue is an insulator, use only a reasonable amount to hold things together; do not cover up the LEDs to the point of concealment. Now all that is needed to form a miniature LED assembly (but not a LED buffer) is a bank of current-limiting resistors. Two of these assemblies can be used to monitor the outputs of the 74154, and portions of one of them can be used to monitor outputs of the experimental circuits. Standard LEDs can also be used, but their bulk prevents producing a compact package.

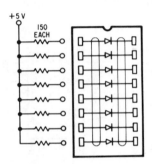

Fig. 10-5 Miniature LED assembly

The 74139

Another type of decoder is the 74139, which is a dual 1:4 decoder in a single 16-pin package. Figure 10-6 gives the pin-outs and the working set-up for the 74139. There are times when two 1:4 decoders are needed. Two 7442s

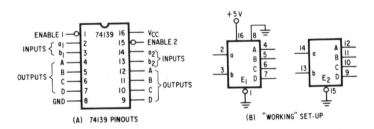

Fig. 10-6 The 74139 dual 1:4 decoder

could be used as a substitute, but using the 74139 eliminates one IC package and is therefore a better solution. Most of the time, it is a question of what is available. If two 7442s are on hand and a 74139 is not, then you should use the two 7442s rather than go out to buy a 74139.

The Seven-Segment Decoders

Figure 10-7 illustrates the common seven-segment decoders. Figure 10-7(a) shows the pin-outs for the 7446 , 7447, and 7448. The 7446 is a seven-segment decoder with high voltage outputs. The 7447 is a seven-segment decoder with lower voltage outputs. Both decoders have outputs that go low with the input binary code. The 7448 is a seven-segment decoder whose outputs go high with the input binary code. The 7446/7447 is used with seven-segment readouts of the common anode type, whereas the 7448 is used with seven-segment readouts of the common cathode type. All these decoders use a 16-pin package.

The 7449 in Fig. 10-7(b) is a common-cathode decoder in a 14-pin package. The lamp test and the BI/RBO inputs have been eliminated; just the BI input is provided.

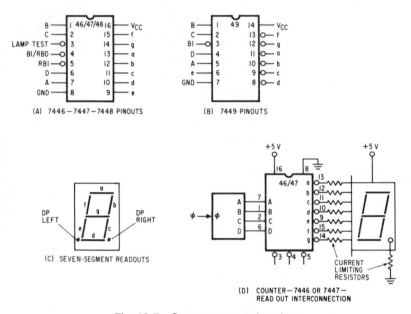

Fig. 10-7   Seven-segment decoders

The seven-segment readout is depicted in Fig. 10-7(c). The segments, identified by lower-case letters, are numbered clockwise, starting with segment "a." This representation is viewed from the top of the readout. Seven-segment displays are normally supplied with a decimal point (DP). They may have either a right-hand or left-hand decimal point, or both, or, in some cases, neither.

Figure 10-7(d) shows how the 7446 or 7447 is connected to drive a common-anode type of seven-segment readout. The counter may be any of those already discussed. If it is a binary counter, then the 7446/7447 will display some weird "numbers" beyond 9. Seven-segment displays are designed to be driven by decimal counters.

In an earlier chapter, we learned how to program a PROM to make a hexadecimal seven-segment decoder. There are seven-segment decoders available to provide the decoding for all 16 input binary states.

The three inputs—Lamp Test, BI/RBO, and RBI—should now be investigated. Lamp Test turns on all the output segments regardless of the inputs on the A, B, C, and D inputs. The number 8 will be displayed.

RBI turns off the output segments, blanking the display. BI/RBO is an input used to suppress unwanted digits in a group of seven-segment displays. If eight seven-segment displays are used in a calculator, the number 00000001 is easier to read when all leading zeros are turned off. For the breadboard experiments, these inputs may be left floating.

Since the seven-segment display is made with LEDs, the LEDs must be current limited to prevent excessive current. Seven resistors are needed for each display used. One resistor is sometimes used in series with the common lead, but this practice results in uneven lighting of the display. When the numeral 8 is displayed, the current drain is much greater than when the numeral 1 is displayed.

The value of the current-limiting resistors is selected to provide the correct lighting level for the display used. This value will range between about 47 and 1000 ohms.

If you come across a circuit using the 7446/7447/7448 /7449 that does not employ any current-limiting resistors, then you should find that the RBI or BI input is connected. If this input is pulsed, the display may be blanked (turned off) part of the time, thus controlling the amount of current through the LEDs that make up the segments. By clocking this input with a clock, the display may be blanked and the current-limiting resistors eliminated. There is one problem. If the clock that pulses the BI inputs on the decoders ever stops, the current-limiting feature may fail and all seven-segment readouts may be destroyed.

Inputs and Outputs

We have been defying convention by using lower-case letters for inputs to the chips and upper-case letters for outputs—a, b, c, and d for the inputs; A, B, C, and D for the outputs—because this seemed to be the simplest arrangement. When lower-case letters are used for the segments in the seven-segment readouts, however, it becomes necessary to assign the capital letters for the inputs to the seven-segment decoder chips. At this stage of your progress, the switch should not be an insurmountable problem.

We have also tried to place the pin numbers inside the chip when giving the pin-outs so that sufficient room will remain outside to note the appropriate pin functions. When working diagrams have been called for, we placed the pin numbers outside the rectangle and the abbreviated function inside the rectangle. This policy should not have produced confusion prior to this chapter. With the 74154, however, we now have one set of numbers inside the rectangle and a second set of numbers outside. It is to be hoped that your knowledge has progressed far enough for you to be able to determine which set of digits refers to what.

There are a great many other decoders that will not be discussed here. The 8000 series of chips include several similar to those of the 7400 series. There are also a number of newer chips designed to provide greater circuit flexibility.

Test Time

The conventional written test used in the classroom to measure student progress is not particularly appropriate for the course outlined in this text. An acceptable substitute might be a "hands on" test. Figure 10-8 shows one such possibility.

As this figure suggests, decoders may be used to simulate the display of a roulette wheel. Fig. 10-8(a) shows a suggested mounting scheme for the LEDs. (Arranging the LEDs in a linear fashion would be more convenient

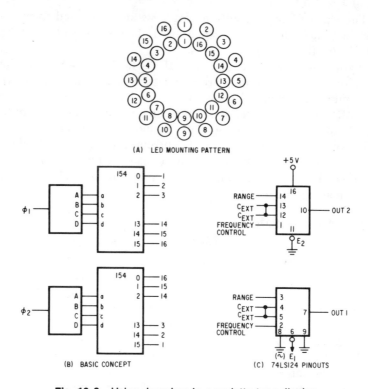

(A) LED MOUNTING PATTERN

(B) BASIC CONCEPT

(C) 74LS124 PINOUTS

Fig. 10-8 Using decoders in a roulette-type display

and just as satisfactory.) Two banks of LEDs are shown, but a single bank would make things simpler (and less expensive). Figure 10-8(b) suggests that one group of LEDs should travel counterclockwise and the other group, clockwise.

Students are bound to ask how the clock can be made to slow down to simulate the true action of the roulette wheel. One possible way to make it do so is with a VCO, or voltage-controlled oscillator. The pin-outs for one VCO are shown in Fig. 10-8(c).

All the other circuits in this text have actually been constructed, and most of them several hundred times, but not the one for this project. It is strictly a circuit for students. Since no information on the VCO is given, it will be up to you to get the chip, look it up in the data manual, get it on the breadboard, and find out how it works. You must then make it do what you want it to do in the circuit. If you can make your wheel perform without being driven up the wall in the process, you will indeed have earned your "A."

Summary

Decoders are a group of chips that accept a binary code input and output a different code. Special decoder chips are needed for special output codes. Any of the 1:X decoder chips may be operated with less than the maximum number of decoded outputs provided that the minimum number of input binary states is used.

For simple decoding, gates are used; for complex decoding, one of the special IC decoding chips.

Chapter 11

Multiplexers
and Demultiplexers

The word *multiplex*—often abbreviated as MUX—means "time shared" or
"time dependent." The time-shared or time-dependent aspect of digital
signals makes the way multiplexers and demultiplexers operate somewhat
difficult to understand.

The 74157

Let us look first at the 74157. This IC is a quadruple two-input data
selector/multiplexer. Its pin-outs are given in Fig. 11-1(a) and its working
diagram in Fig. 11-1(b). The 74157 has two control signals, a strobe and a
select. With the strobe high, all four outputs are forced low. For our first
experimental test on the 74157, let the strobe float, assuming a logic high on
pin 15, and check the four outputs. All should be low.

Now tie the strobe line low with a jumper to ground. The select line
will then determine whether the group of A inputs or the group of B inputs
will be fed to the outputs. Tie all the B inputs low, and let the A inputs float.

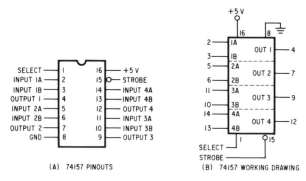

(A) 74157 PINOUTS   (B) 74157 WORKING DRAWING

Fig. 11-1 The 74157 multiplexer

118

With the select line low, the A inputs are selected. Since these are all floating at a logic high, all four outputs should be high. With the select line high, the B inputs are selected. Since these are all tied low, all four outputs should now go low.

This is the test circuit for the 74157, but it really leaves a lot to the imagination about how it works and how it is used. Figure 11-2 is an attempt to transfer the 74157 function to more familiar ground. It shows four toggle switches that select data from channel A or channel B and feed this data to the output channel. Each toggle switch—known as a *4-pole, 2-position* switch—is operated with a single toggle lever. With the handle up, the data from channel A is fed to the outputs. With the handle down, the data from channel B is fed to the outputs. The select signal is the toggle switch handle. The strobe signal is an enable signal that allows the select signal to select either the A channel or the B channel. The mechanical analogy of an electronic 4-pole, 2-position switch is slow and subject to switch contact bounce; the digital circuit is fast, with no switch contact bounce.

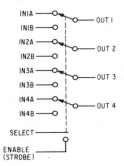

Fig. 11-2 Mechanical equivalent of the 74157

Three-State Devices

TTL chips like the 7400 have what is called totem pole outputs. Two transistors in the output stage of the gate are connected in series. One or the other of these two transistors may be turned on by the gate inputs. If the upper transistor of the totem pole pair is turned on, the output is clamped to the + 5 V supply. If the lower transistor of the totem pole pair is turned on, the output is clamped to ground. Both output transistors cannot be turned on simultaneously because to do so would effectively connect the + 5 V supply directly to ground through the output stage of the 7400.

What this really means is that we cannot connect totem pole outputs of TTL chips together, except by means of additional gating. To allow TTL outputs to be tied together, the TTL group of open-collector-output ICs was developed. The final transistor in the output stage of the gate has an open

collector and a pull-up resistor is added to complete the current path to +
5 V. Therefore, when we need to tie TTL outputs together, we can use open-
collector ICs.

Three-state devices are a relatively recent addition to the TTL family
of ICs. They form a special subgroup of TTL chips with TTL high outputs,
TTL low outputs, and *another* state as well. These chips also have an enable
line. When this enable line is disabled, the outputs of the three-state device
are floated. They are neither high nor low but assume a high-impedance
state. For all practical purposes, the chip is removed from the circuit when
the disable line is placed in the disable state.

An example of the three-state device is the 74125. The pin-outs for the
74125 are given in Fig. 11-3(a) and the working drawing in Fig. 11-3(b). This
chip is a quad bus buffer gate, with active low enables. When the enable pins
are low, it functions as a TTL buffer. Highs in produce highs out, and lows

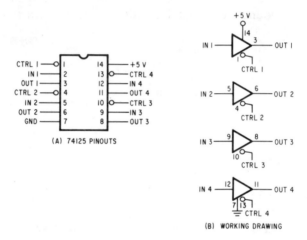

Fig. 11-3   The three-state buffer

in produce lows out. When the enable line is high, however, the chip enters
the third state. The effect is as if the chip were removed from the circuit. It
appears to vanish as far as the rest of the circuitry is concerned.

The four buffer sections behave like four SPST toggle switches to the
rest of the circuit. When the toggle switches are closed, the signals are there
—either highs or lows—but when the switch is open, neither a high nor a
low remains. The toggle switch analogy is presented in Fig. 11-4(a), in which
the four sections of the 74125 are represented as four SPST toggleswitches.
With the switches closed, the signals are transferred to the buss. With the
switches open, nothing loads the buss.

Figure 11-4(b) shows four sections of the 74157 represented as four
DPST toggle switches followed by a 74125 represented as four SPST toggle

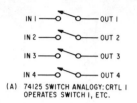

(A) 74125 SWITCH ANALOGY: CRTL I
OPERATES SWITCH I, ETC.

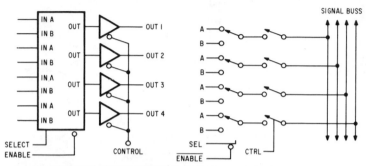

(B) USING THREE STATE DEVICES FOR ISOLATION OF A DEVICE FROM A SIGNAL BUSS

Fig. 11-4 Mechanical analogies

switches. The select line of the 74157 acts as the handle of the 74157 toggle switches. When the 74157 strobe (enable) line is low, this select line can select either the group of A signals or the group of B signals and transfer them to the 74157 outputs. The 74125 can now either place these signals on the buss or completely isolate them from the buss, depending on its enable line. The effect is the same as either closing or opening the four SPST toggle switches. As before, the mechanical analogy of these switches is slow and subject to contact bounce. We shall learn later just how valuable the three-state devices are in the realm of the microcomputer.

The circuits of Fig. 11-4 are drawn in the conventional fashion. Since you now know that each 74125 section is individually controlled by the enable line, we will adopt a slightly different convention for drawing the enable line of three-state devices in the rest of this text. Figure 11-5 illustrates this convention and introduces another three-state buffer—the 8T97. This buffer is a hex buffer in a 16-pin package with two enable lines. One enable line controls four buffers inside the chip, and the other controls two buffers inside the chip. Except for the common enable lines for different sections of the chip, this three-state device works just like the 74125. The 8097, the 74367, and the 74LS367 are all hex three-state buffers similar to the 8T97.

The 74150

The 74150 MUX is a 16-input multiplexer in a 24-pin package. The pin-outs are given in Fig. 11-6. Testing this chip is best done with a more

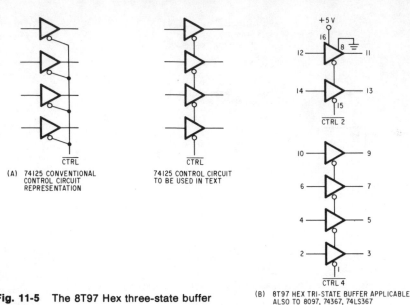

(A) 74125 CONVENTIONAL CONTROL CIRCUIT REPRESENTATION

74125 CONTROL CIRCUIT TO BE USED IN TEXT

Fig. 11-5   The 8T97 Hex three-state buffer

(B)   8T97 HEX TRI-STATE BUFFER APPLICABLE ALSO TO 8097, 74367, 74LS367

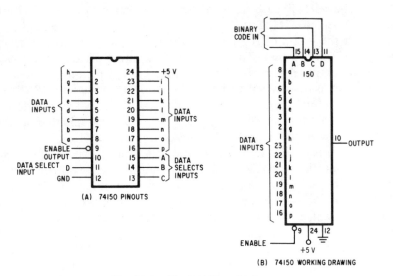

(A)   74150 PINOUTS

(B)   74150 WORKING DRAWING

Fig. 11-6   The 74150 multiplexer

elaborate test circuit. Figure 11-7 repeats an earlier circuit that is also the test circuit for the 74150. (It was used earlier to demonstrate the use of the 74161 as a storage register.)

A clock drives a counter. Here the top 74161 is operating as a binary counter. The binary output of the counter feeds two places: the inputs to the

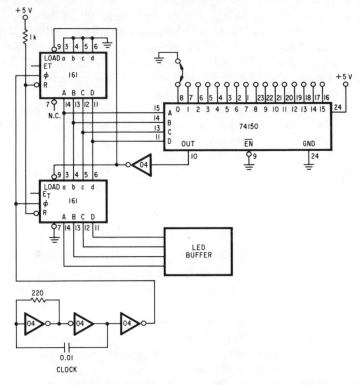

Fig. 11-7 Hexadecimal keyboard encoder

74150 and the inputs to the lower 74161. The lower 74161 is operated as a storage register, not as a binary counter.

Suppose that the binary output of the counter is ∅∅∅∅. The 74150 will "look at" its first input (pin 8, input zero) to see if it is low. If this input is low, the output (pin 10) will go high provided that the strobe (enable) is low. If pin 8 is not low when the input code is ∅∅∅∅, nothing happens to pin 10. As the counter advances, changing the binary code on the input to the 74150, the 74150 is caused to scan its 16 inputs one by one to see if any of them have been taken low. Each of these inputs corresponds to an input binary code. At the time the input is closed, the corresponding binary code also appears on the inputs to the 74150. This same binary code is also applied to the input to the storage register. By taking the high output produced at pin 10 on the 74150 and inverting it, we can take this low and feed it to the load pin of the 74161.

The binary code then present at the preset inputs to the 74161 will now be loaded into the 74161 storage register and appear on its outputs. Moreover, this code at its outputs will not change until a different key is depressed on the inputs to the 74150, and the corresponding binary code for this key is loaded into the 74161. The binary equivalent of the closed input switch is latched into the 74161 by the inverted output of the 74150. By connecting four LEDs or the LED register to the outputs of the 74161 storage register, the code for each switch may be read out on the LEDs. Not only is this a test circuit for the 74150, it is a keyboard encoder circuit as well.

The circuit is also a priority encoder. If two or more switches are closed, the 74150 will "see" only the first closure because the scanning process proceeds sequentially from input zero through input 15.

The 74150 is tested in this circuit simply by closing each input to ground in sequence. The corresponding binary code multiplexed into the 74161 is then read out on the LEDs attached to the outputs of the 74161 storage register.

Repair Costs, Replacement, and Recycling

Electronic equipment is one of the very few items in today's world that is moving opposite to the economic trend. While inflation forces everything upward, electronic prices continue to decrease. The hand-held calculator is an excellent example. Just a very short time back, it cost considerably more than $100.00. When the price dropped below $10.00, it became another of those "beyond economical repair" items. In other words, we have reached the age of "throw-away electronics." The vest pocket transistor radio is another excellent example of the trend.

The average American will put out 10 percent of the initial cost of a piece of equipment for its repair without balking too much. There are certain circumstances under which he will even stand a repair bill of 50 percent. When the cost of replacement exceeds this level, he will invariably opt to replace the unit. You should hold onto such units, however; they can provide you with a valuable supply of parts.

Multiplexing Read-outs

Generally speaking, multiplexing read-outs below a certain number of read-outs is not economically practical for the student or the home experimenter. This number of read-outs might by 6, 8, or even 10.

The read-outs used in calculators are almost always multiplexed. To use them for other than their intended purposes means that we must use multiplexing as well. In the calculator type of read-out, all internal "a" segments are connected together as are all "b" segments, all "c" segments, etc. The individual digits of the display are each provided with a digit enable line. Enabling a particular digit enable line will cause that particular digit to turn on. Whatever is on the segment lines will be displayed by that digit.

If you enter a 7 (with segments "a," "b," and "c," enabled) for example, and then enable one of the digit lines, the corresponding digit will display a 7. If you now enable a different digit, it also will display the 7 on the segment enable lines. So will any other digit enabled by its digit enable line. You might wonder how this type of display can provide different digits in each of its digit display positions. The answer is that we have to multiplex the display.

This type of display comes in two basic varieties. One type is called a *common-anode display* and the other type, a *common-cathode display*. To enable the common-anode display, all segments are activated with a low. The other side of the LEDs that make up the segments are returned to + 5 V. The digit enables are usually just the opposite type of enable; otherwise, they would require an active high.

Figure 11-8 shows a circuit that uses discrete seven-segment LED displays to display the address lines and the data for a computer. This is not a multiplexing circuit, but it will provide us with a starting point. The 8223 hexadecimal decoder fabricated when we experimented with ROMs will be used to provide the segment enable pattern that will cause the display of the numerals 0 through 9 and the letters A, b, C, d, E, and F. The lower-case

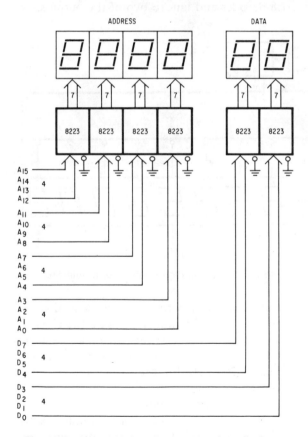

Fig. 11-8 Nonmultiplexed computer data display

letters are used to avoid confusion between the numeral 8 and the letter B, which appear identical on the seven-segment display, and the numeral 0 and the letter D, which also appear identical. Lower-case letters b and d solve the problem only partially because the numeral 6 and the letter b look the same. This difficulty is circumvented by using the a segment for the 6, but not for the b.

A binary code applied to each of the 8223 decoders in Fig. 11-8 will cause the decoded segment to be illuminated on the display so that we can read the decoded numeral or letter. To use a calculator type of display, however, we must multiplex. Figure 11-9 gets us started. If we take a clock, a counter, and a decoder, we can get enables for the digit enables in the calculator type of display. The clock drives the counter to generate a binary code. The outputs of the decoder will go low one at a time, depending on the particular binary input code fed to the decoder at any particular time. We need six of these enables to provide six digits for displaying addresses and data in a computer: four digits for the addresses and two digits for the data. We can use a 1:8 decoder and ignore two of the outputs.

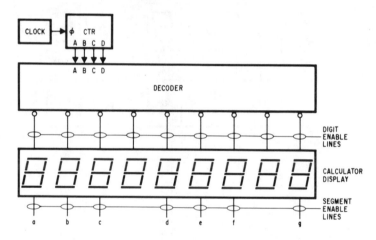

Fig. 11-9   Generating digit enables for a multiplexed display

Set this circuit up initially with a slow clock so that you can see each digit turn on one at a time. Gradually increase the clock frequency until persistence of vision tells you that all digits are on at the same time. You can then replace the slow clock with the clock shown in Fig. 11-10. This clock runs at approximately 200 kHz for the values shown, and all digits thus appear to be on simultaneously, each displaying the same number since we have not multiplexed the segment lines as yet.

If the particular display that you are using requires active high enables instead of the active lows that the decoder is producing, then invert the lows out of the decoder with a 7404 hex inverter to provide them.

On the 8223s in Fig. 11-10 we have a $\overline{\text{CE}}$ pin. The outputs of the 8223s are active only when the $\overline{\text{CE}}$ pin is low. Suppose that we also drive this $\overline{\text{CE}}$ pin with the output of the decoder. Whenever an output of the decoder goes low, one of the 8223s will be enabled and its seven-segment decode pattern will be output. The pattern output will depend on the binary code on the 8223 input pins. Now what do we have? Figure 11-10 shows us that we have a digit enabled and that at the same time we have an 8223 enabled. The output of the 8223 will provide a decode pattern, and this decode pattern

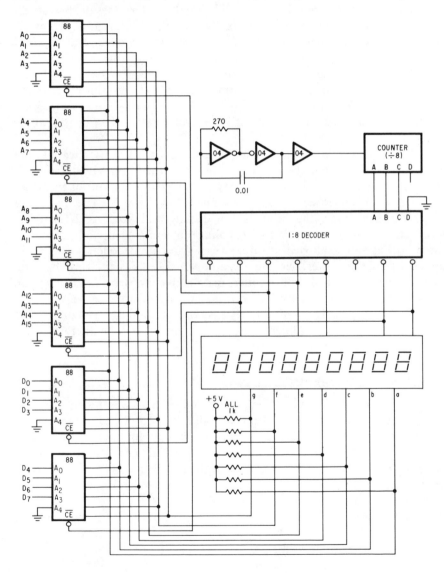

Fig. 11-10 Multiplexing a display

will depend on what binary code is on the 8223 inputs at the time the 8223 is enabled. The digit that is enabled will thus display the input code corresponding to that on the input to the 8223.

A fraction of a second later, this digit is no longer enabled, nor is the 8223, but a different digit is enabled and so is a different 8223. If this 8223 has a different binary code on its input, then a different digit will be displayed by the digit that is enabled at this particular time.

The open collector outputs of the 8223 require seven resistors. These resistors are also shown in Fig. 11-10. Their value determines the brightness of the display. We can now display six digits on the display, and each of the digits displayed can be different. In other words, we have a multiplexed display. This display can be run with a slow clock, but if you use one, be sure to add current-limiting resistors so that you do not destroy the LEDs. If you use a high-speed clock, you can eliminate the current-limiting resistors because the LEDs in the display will be on for only one-sixth or one-eighth of the total time. We pulse them with the circuit rather than operate them statically.

You will also notice that the LEDs have dimmed down. Since they are on only part of the time, they are not as bright as they would be if they were on all the time. To get the same energy (same brightness) out of them, we would have to increase the supply voltage above + 5 V. The higher we increase this voltage, the brighter the display will become. TTL chips, however, have a maximum working voltage of + 5.5 V. If we increase the supply voltage much above this value, we can ruin our decoder. An 8223 will not be affected, however, because it is an open-collector chip. If we have to, we can use an open-collector decoder, or, if active highs are needed, we can use an open-collector hex inverter to follow the decoder instead of the 7404.

This multiplexing circuit is fairly easy to understand. If you object that it requires burning six 8223 decoders, just remember that the discrete circuit also requires six decoders.

Another Multiplexing Circuit

Let us give credit where credit is due. This circuit was designed by one of the author's ninth-grade students who didn't like the idea of burning six 8223s either. It is a better and less expensive circuit than the one it modified. Figure 11-11 shows the original circuit. Its 8T97s also have a $\overline{CE}$ pin. If we drive the 8T97s (or any other member of the three-state buffer family) with the outputs of the decoder, they will transfer whatever is on their inputs to their outputs when the $\overline{CE}$ pin is low. When the $\overline{CE}$ pin is high, they are three-stated and might just as well not even be there. They act just like electronic toggle switches. When the $\overline{CE}$ line is low, the toggle switches close, and the data is fed to the 8223. When the $\overline{CE}$ line goes high, the toggle switches open and, because they are open, have no effect on the 8223 inputs.

The FNA-45 Display

Since we must pulse the LED display in a multiplexed circuit, the display is going to dim. Fairchild makes a display that has a lot of internal circuitry to overcome this problem. The Fairchild FNA-45 is a nine-digit, multiplexed, calculator type of display with numbers ¼ inch high. It makes a very nice computer read-out device.

The FNA-45 requires active highs on both the segment enables and digit enables. We must therefore use a 7404 hex inverter to invert the lows out of the decoder in order to get active high digit enables. The FNA-45 display was designed to operate from + 5 V. Because of its internal circuitry, it does not require current-limiting resistors. The circuit is shown in Fig. 11-11; this is the circuit that we will use in the following chapters. Many other circuits may also be used for multiplexing seven-segment read-outs, but since multiplexing of digital circuits is not an easy thing to understand, we shall not discuss them here.

Demultiplexing

In multiplexing, we *combine* data on a time-coincident basis. When we *demultiplex*, we *separate* data on a time-coincident basis. A decoder is a demultiplexer. A binary code goes into the decoder, and a separated code comes out. Data manuals often refer to a device that performs this function as a decoder/demultiplexer.

Let's combine a multiplexer and a demultiplexer in the same circuit to make a keyboard encoder. One of our earlier circuits would encode 16 switch closures to ground and convert these to a binary code output. This arrangement worked fairly nicely because one side of all switches were common and could be connected to ground, but what happens if the switches are in a matrix and have no common line? How can you make a circuit that will "scan" the switch matrix?

Figure 11-12 illustrates the problem. Here is a switch matrix in which the switches are arranged in columns and rows. The closure of a single switch will connect a vertical line to a horizontal line. How does one determine which switch in the matrix has been closed? The matrix, it must be understood, can have any form. It can be square, such as 8 lines wide and 8 lines high. It can be rectangular, such as 8 lines wide and 4 lines high or 4 lines wide and 8 lines high. It might be 7 X 6 or 9 X 11 since the number of lines may be odd or even.

Figure 11-13 shows a calculator keyboard with a switch matrix that is 4 X 5. We desire to encode this keyboard as a hexadecimal keyboard encoder to use with a computer. The circuit in Fig. 11-13 will do the job. We can use a matrix of 4 X 4 and ignore four of the keys. Since we need only 16 switch positions for a hexadecimal keyboard, a 1:4 decoder/ demultiplexer, 1:4 multiplexer, and a few TTL chips are all that is required.

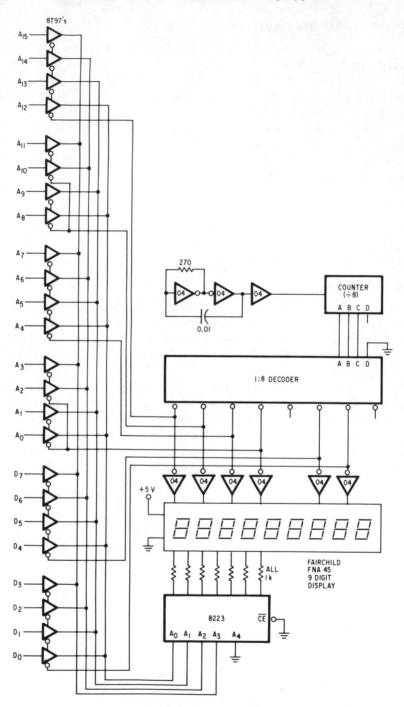

Fig. 11-11 A less expensive and better multiplexed display

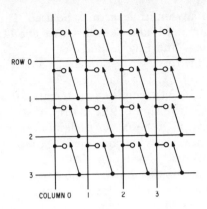

Fig. 11-12 A switch matrix

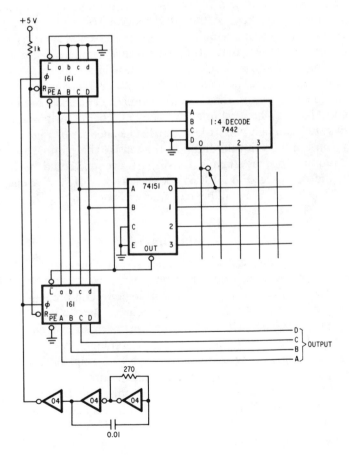

Fig. 11-13 Hexadecimal keyboard encoder for a switch matrix

Any decoder and any multiplexer may be used. The 7442 is operated as a 1:4 decoder by tying the C and D inputs low. The 74151 is operated as a 1:4 MUX by tying its C and Enable line low.

The clock drives the top 74161 counter. Two of its outputs go to the 7442, and two go to the 74151. The ϕ, 1, 2, and 3 outputs of the 7442 go low in sequence, causing the 7442 to scan the switch matrix from left to right and placing lows on each vertical line of the matrix in sequence. The counter inputs to the 74151 cause it to "look at" each input in sequence to see if it is low. After the 7442 goes across the row, the 74151 moves down a row, and the 7442 goes across this row as well. The 74151 then moves down another row, and the 7442 goes across the new row, and so on. When the 74151 finds one of the switches closed, it outputs a low-going pulse and loads the second 74161 storage register. The output lines reflect the binary code equivalent of the switch closed in the matrix.

If you need to encode more than 16 switches, Fig. 11-14 goes to the opposite extreme. By using a 74154 and a 74150, we can have a matrix large enough to encode 256 keys. Although it requires more counters and storage registers, the basic circuit remains the same. Sixteen squared is 256 and so is 2^8. We can generate a byte of data with this encoder. The only problem with entering data into a computer is that it would require 256 switches to do the job—not too practical a method. However, there are times when you will want to encode more than just 16 switches. The basic circuits of Figs. 11-13 and 11-14 can be used to encode any number of switches, such as the 80 to 90 switches on an ASCII keyboard. The encoder of Fig. 11-14 will do the job, but don't use all the lines. If the matrix that you desire to encode doesn't require 16 lines, then a 1:8 or 1:10 decoder could be used or the MUX could be changed to a 1:8 MUX such as the 74151. The encoders will encode the switches. You will have to find ways of handling the shift and the control characters.

Summary

Multiplexing means "time-coincident" or "time-shared." In multi-plexing, we combine data, and in demultiplexing, we separate data. Both are tricky and not the easiest thing in digital logic to comprehend.

The completion of this chapter concludes the study of the ICs that make up the 7400 family. It has by no means been an exhaustive survey. It has been just broad enough to create a sufficient background for the forthcoming chapters. There are many more TTL chips on the market, and more new chips are being manufactured all the time.

Information on ICs is obtained from books called *data manuals* These manuals are offered for sale by all major IC manufacturers, and their price is quite modest. If you continue to experiment and work with ICs, then you will need to supply yourself with one. Data manuals are frequently for sale

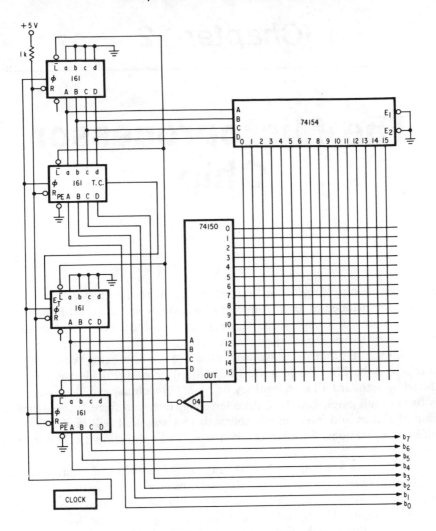

Fig. 11-14 Encoding more switches in a larger matrix

at electronic stores. Examining them before buying is a good move. Unfortunately, not all stores stock them, and not all manuals from all manufacturers are stocked.

Chapter 12

The Microprocessor Chip

Some people may consider it "out of place" to find a chapter on microprocessors included in an introductory electronics textbook. Although not "out of place," it is beyond the scope of this book to go further than some preliminary basics about microprocessors.

All the ICs that you have been studying are now being replaced by a single IC—the LSI (Large Scale Integration) chip. The LSI may be produced by standard TTL technology or by MOS technology. The basic idea is to place all gates, counters, decoders, registers, and so forth on a single chip of silicon and interconnect them all in a way that will facilitate their different operations. The chip is "instructed" what function it is to perform by "feeding" it binary codes.

Only a few years ago, a NAND gate was constructed of vacuum tubes, capacitors, resistors, and diodes. The vacuum tube gave way to the transistor, and transistors, capacitors, resistors, and diodes were then used to make the NAND gate. The NAND gate subsequently gave way to the integrated circuit, and it is IC technology that forms the basis of this textbook. In like manner, computers were first made of vacuum tubes, then transistors, then integrated circuits, and now the latest phase of technology has brought us to the "computer on the chip."

Computers or logic functions based on the older technologies were termed *hard-wired devices.* To change the functions of a hard-wired circuit, the wiring itself had to be changed. When a microprocessor is used, the wiring is not changed to change the logic function; rather, the "instructions" to the microprocessor are changed. Changing the wiring is a major task; changing the "instructions" is a relatively simple one.

Circuits and physical devices that are interconnected (hard-wired) are logically called *hardware.* The "instructions" are called *software* to distinguish them from the hard-wired devices.

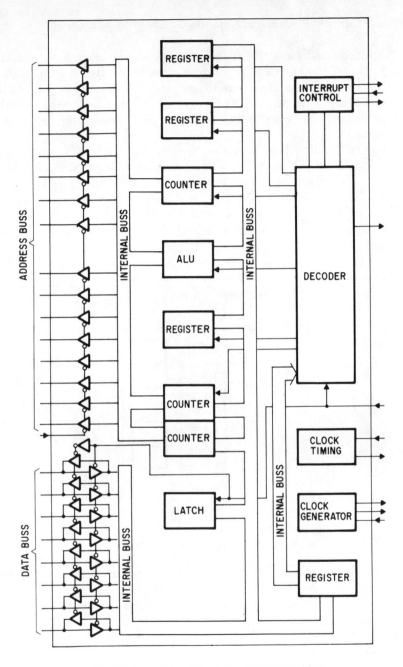

Fig. 12-1 Inside of the microprocessor chip

An inspection of Fig. 12-1 will reveal many of the names of the devices that you have been studying in this text, and almost all of them should now be familiar to you. Figure 12-1 is a block diagram of what is inside a typical microprocessor chip. It contains one block called an *ALU* (an abbreviation for Arithmetic and Logic Unit). All microprocessor chips have one. The TTL ALU of the 7400 series, for example, is the 74181. Its pin-outs are given in a TTL Data Book, and you can investigate its operation on your solderless breadboard. The ALU is the portion of the microprocessor chip that does the adding and subtracting and also performs certain other logic functions.

Figure 12-1 reveals that there are a very great many counters, registers, decoders, and so forth inside the microprocessor package. Since the microprocessor quite often forms the "heart" or "Central Processing Unit" of a computer, it is also termed a *CPU*. Since it is also referred to as the "Master Processing Unit," it is sometimes called an *MPU*.

Figure 12-2 shows the fundamental blocks that are used to make a computer. We must have a power supply to run everything. We must have some kind of memory to store the binary code that informs the CPU of the task we wish it to perform. We must have a way to get data into the computer and out of the computer ("input" and "output").

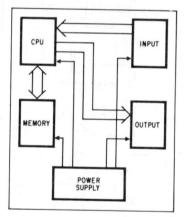

Fig. 12-2, Block diagram of a simple computer

Computers are dumb. You have to tell them every little detail of *what* you want them to do. Then you must tell them every little detail of *how* you want them to do it. These are the function of what is known as the *software.* Once the computer "understands" what it is that you want it to do, and how it is to do it, it will do it very quickly. The person who writes the instructions (software) for the computer is called a *programmer.*

The input and output blocks of Fig. 12-2 provide a means of getting the instructions into the computer and the computer's response back out of the computer. The device that is used to place the instructions in the computer is called the *Input (I).* The device that allows the computer to communicate back to us what it has done with our instructions is called the

Output (O). Since the two devices, Input and Output, must operate hand-in-hand, you will most often see them referred to as *Input/Output,* or simply *I/0*. The term *I/0* may refer to the physical devices that do the inputting and outputting or to the process of inputting and outputting.

Systems

Figure 12-2 showed the basic blocks needed to make a CPU function and do something. Figure 12-3 adds more memory and also different kinds of memory. It adds quite a few blocks called P_1, P_2, P_3, and so forth. These "Ps" are *peripherals*. A peripheral is a device that is added to the basic computer to increase its computing power or to improve the operator's convenience. Peripherals include things like Teletypes®, glass teletypes (or video monitors and electronic keyboards),[1] cassette tape recorders, floppy discs, paper tape readers, and hard-copy devices such as high-speed printers. Since each addition to the basic CPU increases the power requirements, the power supply block in Fig. 12-3 has been enlarged considerably. Since Fig. 12-3 also shows the CPU at the "center" of things, you can readily see why the name CPU has been applied to the microprocessor chip.

[1] A glass teletype is an ASCII keyboard and a video monitor with serial I/0 at 110 band. It provides all the functions of the ASR 33 Teletype without hard copy.

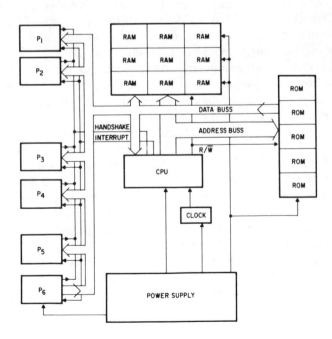

Fig. 12-3   Block diagram of a computer system

When all the components of Fig. 12-3 are interconnected, we have a computer system. It should be fairly obvious that the more memory and peripherals you have, the more powerful your system and the more (and easier) computing you can do. It should also be rather obvious that the more you have in your system, the more money you will have tied up in it.

It is not the purpose of this chapter to discuss the different microprocessors available from different manufacturers, which cover a very wide range of types, bits (word size), power consumption, and price. Since the cost of a particular CPU is only a very small part of the cost of the computer *system,* a choice based on the cost of the CPU chip itself is of minor significance.

Of far greater significance is the quantity of code already written for a CPU. It takes a great deal of time to write all the binary code (software) that will tell the CPU what it is to do. One of the most relevant points in choosing a CPU, then, is to see what software is already available for it. The choice of CPU then becomes more a question of its popularity than it does of the number of instructions in the instruction set or the cost of the microprocessor chip itself.

There is little question that the Intel 8080 and its newer counterpart, the Intel 8085, are the most popular CPUs. Second in popularity is the Motorola 6800. Third is the MOS Technology 6502, and fourth is the Zilog Z80. Any of these CPUs make a good choice. What about the others—the 2650 by Signetics, the 1802 by RCA, and all the rest of the different CPU chips by various manufacturers? They are all good chips and deserve your consideration, but they just didn't catch on.

Making your own computer is no longer economically feasible. The manufacturers of the simpler machines have the potential of producing an operable computer at about the same price that we would have to pay for the parts alone to make our own machine.

If you reflect for only a moment about what happened in the small calculator field, you will see that it is exactly what has happened in the computer field. The initial price of the hand-held calculator was above $200. As technology advanced, the price of the calculator dropped. Eventually the price of a functioning calculator dropped below the price of the individual parts used to manufacture a home calculator. This drop happened over a period of four to five years. In the computer field, the pace accelerated, and the same thing has already happened. The only logical reason to be advanced for making your own machine is the educational aspect. To understand how the next generation of electronics is going to relate to you and your work, there is no better way to obtain this knowledge than to homebrew your own computer. But remember, it is no longer economically feasible to homebrew your own machine.

Chapter 13

Power Supplies

Of all the aspects of electronics, the power supply is the simplest and the most fundamental. And yet, it causes more problems than any other portion of the circuitry.

Every piece of electronic equipment requires a power source of some form for its operation. This may be a battery. Unfortunately, batteries have the nasty habit of being consumed as they are used and going "dead" at precisely the wrong times. Although batteries provide a simple source of power for many applications, they are not inexpensive. Consequently, the power required to operate electronic equipment is usually derived from ac mains with some hardware that we call the *power supply*.

DC from AC

The purpose of the power supply is to take the alternating current (ac) from the power supply mains and change it to direct current (dc). The nominal 120 V available from the ac outlet is transformed with a *transformer* to a lower voltage. This transformation is accomplished by the turns ratio of the windings in the transformer. If the transformer contains many turns of wire in the primary connected to the ac mains and fewer turns for the output windings the voltage will be transformed downward, or "stepped down." This downward transformation is accompanied by an increase in current. As the voltage is transformed downward, the available current is transformed upward because the power into the transformer is essentially the same as the power out of the transformer, and power is the product of the voltage and the amperage.

If the transformer has 200 turns for the primary winding connected to the incoming 120 VAC and the secondary has 20 turns for the outgoing ac, the turns ratio is 10:1 and the output voltage will be one-tenth the input voltage; thus the secondary will supply about 12 V. If the current drawn by the primary winding is 1 A, then the 10:1 turns ratio will provide an output current of 12 A in the secondary, or output, winding. Since more current is

139

available from the secondary, the wires used for the windings of the secondary will be physically larger than the wires used to wind the primary.

The power available from a transformer is a function of the amount of iron used to fabricate the transformer. A physically small transformer will handle less power than one physically larger.

The input to a transformer must be ac, and the output must also be ac. The transformer will burn up if dc is applied to the windings. The output of the secondary of the transformer is changed to pulsating dc with *rectifiers*. A diode is a device that allows current to pass through it in one direction only; consequently, a diode is a rectifier. Since the pulses from the diodes are of one polarity only, the diodes convert the transformer ac to pulsating dc. A *filter capacitor* is used to store the pulses. The filter capacitor charges up on each pulse and then discharges as the dc is delivered to the circuit that the power supply is feeding.

The circuit that the power supply feeds is called the *load*. The load on the power supply is represented as a resistor. It is not actually a resistor, but the circuitry that is using the power supplied by the power supply. To the power supply, this circuitry appears to be a resistor; thus it has become the convention to depict these circuits as a resistor and call the circuitry the *load* on the power supply.

Since electrons are stored in the filter capacitor in pulses, the larger this capacitor, the more electrons it can store and the smoother the dc supplied to the load will be. The fluctuating dc out of the filter capacitor is called *ripple*. The better the power supply and the larger the filter capacitor, the less ripple and the smoother the dc.

Circuit Configurations

Figure 13-1 shows the simplest power supply. This circuit, which uses only one diode, is called a *half-wave rectifier*. The diode converts only one-half of the ac cycle to a dc pulse. It is used only for the simplest power requirements.

Figure 13-2 shows a full-wave, center-tapped circuit. Two diodes are used with a center-tapped winding on a power transformer. Both halves of the ac cycle are converted to pulses to allow the filter capacitor to charge up

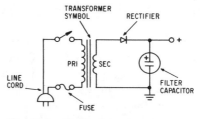

Fig. 13-1 The half-wave power supply

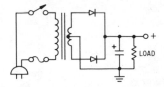

Fig. 13-2 The full-wave, center-tapped power supply

twice as often so that there will be less ripple and smoother dc out of the power supply. The pulses to the filter occur at 120 times per second whereas the mains produces 60 cycles per second.

Figure 13-3 shows a full-wave bridge circuit. Four diodes are used, and a pair of them conducts on each half of the ac cycle. This circuit does not use a center-tapped transformer but still provides an output dc pulse for each half of the incoming ac so that it is also a full-wave rectifier circuit and provides 120 pulses per second out for 60 cycles per second from the mains.

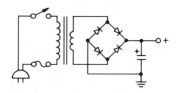

Fig. 13-3 The full-wave bridge power supply

Regulation

The output voltages of the circuits of Figs. 13-1, 13-2, and 13-3 will decrease as the load on the circuit increases. As more electrons are drawn off the filter capacitor by an increasing current through the load; the output voltage sags. The larger the filter capacitor in the circuit, the less this sag will prove to be.

Electronic circuits do weird things when their operating voltages change. To prevent these undesirable changes, voltage regulation is used. The circuitry used to provide it can be quite extensive. One of the boons accompanying the computer revolution has been the development of the IC voltage regulator. Although these handy devices have large quantities of electronic circuitry in them, they have only three terminals and look like power transistors. The three terminals are labeled In, Out, and Common. The regulators are manufactured to provide a fixed output voltage at a given current rating. Figure 13-4(a) shows the more common regulators and their pin-outs, and Fig. 13-4(b) indicates the internal circuitry.

Zener diodes are another form of voltage regulators. Called *shunt regulators,* they are used to provide a regulated voltage when the amount of

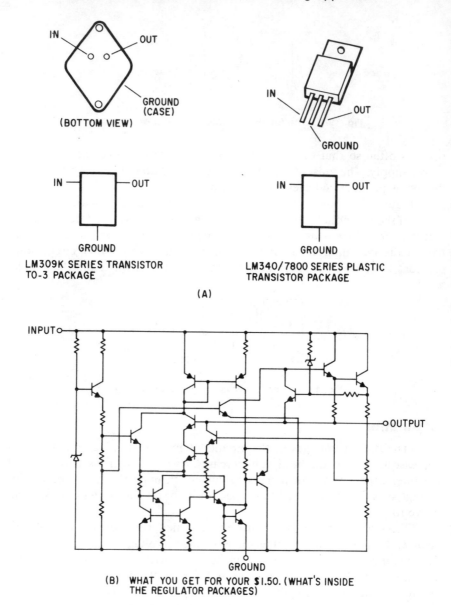

IN
OUT
GROUND
(CASE)
(BOTTOM VIEW)

IN
OUT
GROUND

IN —— OUT
GROUND
LM309K SERIES TRANSISTOR
TO-3 PACKAGE

IN —— OUT
GROUND
LM340/7800 SERIES PLASTIC
TRANSISTOR PACKAGE

(A)

INPUT

OUTPUT

GROUND
(B) WHAT YOU GET FOR YOUR $1.50. (WHAT'S INSIDE
THE REGULATOR PACKAGES)

Fig. 13-4 Three-terminal voltage regulators

current required is quite small. Like all diodes, the Zener diode requires use of a current-limiting resistor to prevent it from drawing excessive current.

We have already learned that a conducting silicon diode has a fixed voltage across it of about 0.6 V and that a conducting germanium diode has a fixed voltage across it of about 0.2 V. This voltage, which is constant, can

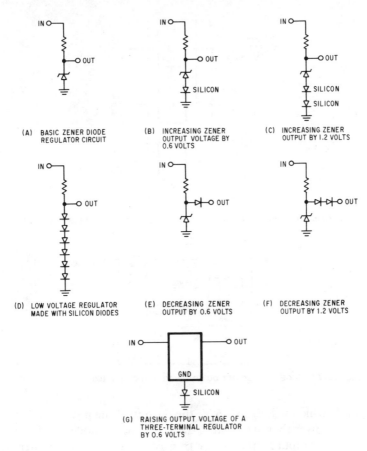

(A) BASIC ZENER DIODE
REGULATOR CIRCUIT

(B) INCREASING ZENER
OUTPUT VOLTAGE BY
0.6 VOLTS

(C) INCREASING ZENER
OUTPUT BY 1.2 VOLTS

(D) LOW VOLTAGE REGULATOR
MADE WITH SILICON DIODES

(E) DECREASING ZENER
OUTPUT BY 0.6 VOLTS

(F) DECREASING ZENER
OUTPUT BY 1.2 VOLTS

(G) RAISING OUTPUT VOLTAGE OF A
THREE-TERMINAL REGULATOR
BY 0.6 VOLTS

Fig. 13-5   Zener diode regulators and silicon diodes used for voltage regulation

be used in place of Zeners, or in conjunction with Zeners, to form fixed voltages for regulating purposes. Figure 13-5 shows some of the circuit configurations that can be used with ordinary diodes to provide Zener type regulation or to make slight alterations in a particular Zener output voltage.

Boosting Current

The available three-terminal regulators have capacities of 1 A, 2 A, and up to 5 A. Additional current demands from a circuit are met by using the regulators in parallel or by using a power transistor. The basic circuit is that of the emitter-follower; the transistor used in the circuit for this purpose is called a *series pass transistor*. Figure 13-6 shows how additional current requirements can be met by paralleling three-terminal regulators. Figure 13-7 shows how the power transistor may be used to boost the current available from the regulator circuitry. The BE junction of the power transis-

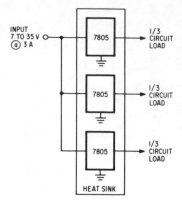

Fig. 13-6   Paralleling three terminal regulators in increase current

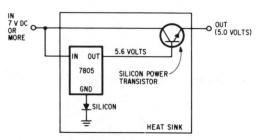

Fig. 13-7   Boosting current output with a series-pass power transistor

tor is a diode which will drop 0.6 V; the output of the power transistor will thus be 0.6 V lower than we want it to be. A silicon-diode regulator can be used to raise the input to the power transistor by 0.6 V to compensate for this drop, as also shown in Fig. 13-7.

Positive and Negative

Computers often require both positive and negative voltages for their operation. "Positive" and "negative" refers to some reference point. If you examine the circuit of Fig. 13-8, which is drawn without a ground or common reference terminal, you will see that the outputs of the power supply are marked simply "+" and "–". In Fig. 13-9 we have grounded the

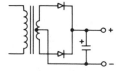

Fig. 13-8   The "unreferenced" power supply

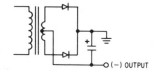

Fig. 13-9   Negative voltage output with reference to ground

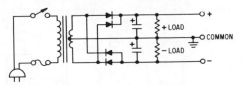

Fig. 13-10 Producing both positive and negative outputs from the same transformer

plus output of the power supply. The minus terminal can now supply a negative voltage, but it is still the same power supply! The only thing changed is the addition of a *reference point*—the ground.

Figure 13-10 shows the full-wave center-tapped circuit redrawn in a different configuration to provide both plus and minus outputs with reference to a common point, the ground. Negative three-terminal regulators are available as well as positive three-terminal regulators; their use is depicted in Fig. 13-11.

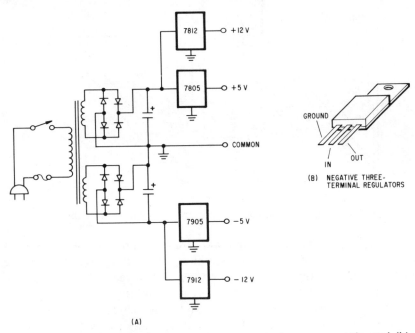

Fig. 13-11 (a) Four different regulated outputs from one supply, and (b) negative three-terminal voltage regulator and its connections

The Power Supplies

Figure 13-12 shows a power supply suitable for most experiments on a breadboard.

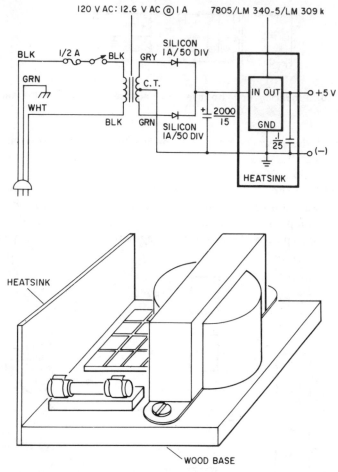

Fig. 13-12   Construction of a simple power supply for the console

Until recently, power supplies were built on a metal chassis. With an ever-increasing number of power supplies being constructed, a simpler method of construction evolved during the past two years. A block of wood for a base and a piece of metal to provide heat-sinking now takes the place of the metal chassis. The transformer and other components are simply fastened to the wooden base. A handy way to make tie points is to take a piece of circuit board and with a hacksaw cut just through the foil. By making cuts at right angles, the copper foil can be formed into islands, and these islands can than be used to make all soldered connections. It is a fast and very inexpensive method. One or two holes drilled in the corners allow the PC board to be secured to the wooden base. A small piece of metal or scrap PC board is added to the wooden base to make a heat sink.

In the circuits of Figs. 13-9 and 13-10, the potential of the heat sink cannot be overlooked. If the heat sink is electrically tied to the chassis or frame, then it will be at "common" potential. If a regulator IC or a transistor used as a regulator element is bolted to the heat sink, the case of the IC or the transistor must either be at the same potential as the heat sink or must be electrically isolated from it. Mica insulation kits are available for this purpose and must be used if the case of the regulator element and the heat sink are at different potentials.

Figure 13-13 shows a power supply developed for extensive experimentation with solderless breadboards. A fixed + 5 V is available for

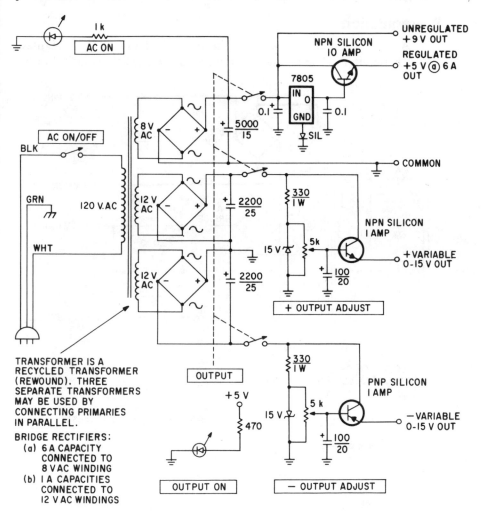

Fig. 13-13 A multipurpose experimenter's power supply

work with TTL circuits. Variable voltages, which can be set with potentiometers, are available for other experimental purposes. One switch discon-
nects all power to the output terminals.

Figure 13-14 shows a computer power supply. The outputs are all
unregulated, and the regulation is accomplished on each board in the system
with three-terminal regulators. The relay in the off-on section can be replaced with an ordinary toggle switch. We use a relay because if the power
should ever be lost from the ac mains when we are not around, the computer
will not come back on until we turn it on. It is a very inexpensive safety
feature in comparison to the total investment in the computer system.

Preregulation

A voltage difference between the input and output of a regulator
circuit is required if the regulator circuit is to accomplish its task. For a +

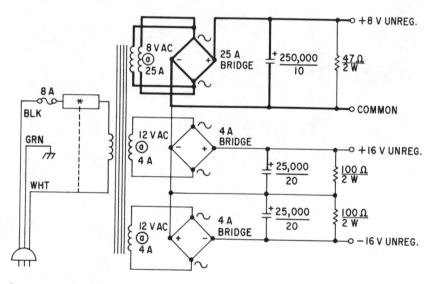

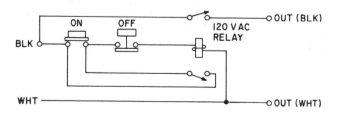

Fig. 13-14 A computer power supply

5-V three-terminal regulator, at least 7 V should be input. Such regulators will work with only a 1-V margin, but if the ripple out of the supply should increase, this ripple may get into the computer and raise havoc.

On the other hand, the input to three-terminal regulators can be as high as 35 V. The problem is that all unnecessary extra voltage is turned into heat in the regulator. The regulator generates heat just doing its job, and the additional heat generated by the extra input voltage means that less usable output is available for the load.

If the unregulated voltage into three-terminal regulators is higher than 7 to 8 V, then a preregulator circuit can be used to take the unneeded voltage (and heat) away from the regulator. Such a circuit is shown in Fig. 13-15. A

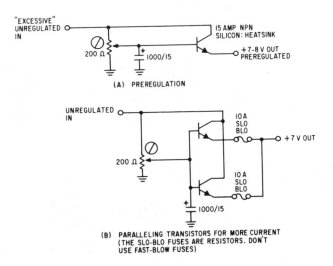

(A) PREREGULATION

(B) PARALLELING TRANSISTORS FOR MORE CURRENT
(THE SLO-BLO FUSES ARE RESISTORS. DON'T
USE FAST-BLOW FUSES)

Fig. 13-15   Preregulator circuits

power transistor (or even several) is used to preregulate the + 8 output line. The preregulator also reduces the ripple content of the + 8 line so that the on-board three-terminal regulators can be operated with an input voltage of about + 7 V. This practice allows much cooler operation of the on-board regulators in the densely packed circuit cards. The preregulator transistor is located on the main-power-supply heat sink, which is massive and near the exhaust fan so that the heat generated will be carried away from the circuit cards.

This is not a "total cure all" circuit. Each time a new circuit card is installed in the computer, the additional current drawn causes the un-regulated (or preregulated) + 8 line to sag, necessitating resetting the control pot in the preregulator circuit to allow the preregulator to supply the additional current demands.

Summary

Power supplies are the simplest of all electronic circuits. They are also the most frequently encountered circuit because every piece of electronic equipment requires one for its operation. Power supplies also require more troubleshooting than other electronic circuits, probably because they are more common and because they must function correctly before anything else will work.

Chapter 14

Troubleshooting

It is unfortunate that a chapter on troubleshooting needs to be included since it implies that you are going to have some problems. Well, since you will definitely have some problems, we might as well face up to them.

The Tools

The most fundamental aid in troubleshooting is your brain. Stored away in your own memory bank by this time is a considerable amount of knowledge on how some of the ICs function. You must now gather "clues" like a detective and then use logic to "solve the case."

The technicians' first tool is the voltmeter. Electronic circuits function only when the proper voltages are applied. The first step in troubleshooting an inoperative or malfunctioning circuit is to use your voltmeter to verify the presence or absence of voltages at the ICs. A common error is to measure the applied voltage at the regulators or across the filter capacitors in the circuit. This is a valid test in itself and definitely a *part* of the verification of the voltages. However, suppose that one of the pins on an IC got bent under the IC when it was inserted into its socket, or even when the IC was soldered onto the PC board? This has happened even on commercial PC boards and the fault has escaped quality control inspection. What happens if all but one of the pins on an IC or the sockets that hold the IC are missed in the soldering process? The power to the IC can be there initially but "disappear" later and cause the need for troubleshooting.

Voltages to ICs are measured on the top of the board and directly on the pins where they enter the IC package. This verifies the presence or absence of voltage at the last accessible point in the circuit for the technician. The IC connections inside the package can also be broken, but this fault will be manifested only when the IC is tested and shown to be defective.

The voltmeter can be any quality instrument, analog or digital. But how do you know that your voltmeter isn't lying to you? Any voltmeter can get out of adjustment and give erroneous readings. The newcomer to electronics tends to have complete faith in his voltmeter and assumes that his instrument is infallible. Not so! Voltmeters must be calibrated.

To calibrate a voltmeter, its reading must be compared against a "known" voltage. Where does one get this known voltage? An ordinary, fresh flashlight cell has an output voltage of about 1.55 V. It can also have any other voltage near this value depending on how "fresh" it really is. A better reference is the mercury cell. This battery cell has a voltage—1.35 V —that is very constant over the entire life of the cell. Both these standards have been frequently used to give some degree of assurance that a test instrument is functioning correctly. Another way to obtain a calibrating standard is to use a precision Zener in the manufacturer's recommended circuit. The point is that you must not blindly assume the correctness of your voltmeter. If you are going to depend on what it is saying to you, you must have some means of verifying that what it is saying is correct.

Highs And Lows

One of the nicest things about digital electronics is that we are concerned with highs and lows rather than absolute voltages. A high is a voltage up to about + 5 V. Preferably, it should be a voltage around 3.5 V or more. Digital ICs will function with "highs" as low as 2.5 V, but some circuits will not function correctly below this value. When the "high" drops down to 1.5 V, you can expect lots of problems and faulty operation.

A low is any voltage near ground, or 0 V. Anything under a diode drop (0.6 V) will pass for a low. When the "low" approaches 0.8 V, you can begin looking for problems to appear. When it gets as high as 1.0 V, you can expect disaster.

The voltmeter can be used to measure the highs and lows of a digital circuit. A logic probe can also be used to measure them. The simple LED testers and LED buffers that we used throughout the text were all logic probes. A number of digital logic probes have appeared in the literature in the past few years, and we will give you a circuit for one of them in this text. This probe is built into a console to facilitate more rapid experimentation on the breadboard. You can even make a very fancy logic probe with a seven-segment read out so that the indication will be an "H" for a high, an "L" for a low, and a "P" for pulse.

The Oscilloscope

Another tool for troubleshooting digital circuits is the oscilloscope, or scope. This is an expensive tool but indispensible for tracking down some of the problems that can occur. How to use a scope and interpret its display is possibly the most difficult thing to learn in the field of electronics. Many attempts have been made by many different authors to teach the subject. The present author can tell you one important thing from many years of experience and that is that you can learn to use a scope only by using one and that a "resource person" can be of more help to you than any book. I strongly urge you to seek out such a "resource person" and learn what a good scope is like, and how it works before you rush out to buy one. A good scope is a major investment, and you should be sure of your ground before you invest.

Troubleshooting and Homebrew

Troubleshooting implies that the equipment being investigated once worked and has failed and that you must now find what has failed. When a piece of equipment has been homebrewed, it has not yet functioned, and the term "troubleshooting" becomes a misnomer. There is a vast difference in troubleshooting a piece of equipment that was once operational and has failed and trying to find out what is wrong with your own creation that has still to prove itself. The process of getting things operational is not dissimilar from getting things *back* into operation, but it is many times more difficult than the other. When you do *all* the "engineering" and *all* the construction, a great many more sources of error are possible. Even a kit that you assemble is far more likely to be easier to get operational than homebrew equipment. The basic engineering has already been done, and many of the potential bugs have already been eliminated; since it is a kit, moreover, you can assume that it will function when it is correctly assembled. By contrast, homebrew equipment will function only when you yourself get all the design errors out and all the bugs out. Once it is operational and then subsequently fails, the problem becomes one of troubleshooting as with any other piece of electronic equipment. "Debugging" then is the term that we will use to get a piece of equipment operational—or, as the expression has been evolving in the new computer field, "up and running"—whereas "troubleshooting" will be left to the process of finding the problem in a piece of equipment that once was operational and now has failed.

Divide and Conquer

The fundamental principle of troubleshooting is that of splitting a circuit into halves. Let's use an ordinary radio as an example. The first time anybody looks at the underside of an old-fashioned radio he is "floored" by the mass of wires. Buried beneath and tangled among the forest of wires are round things of different shapes and sizes. If the radio is of a later vintage, then the forest of wires has been replaced with a circuit board with little hair-thin traces of copper running around in complex patterns. The other side of the board, however, resembles a forest, with the round things sticking up in a densely packed mass.

Complex as the radio may seem, it is a very simple device when compared to any of the modern digital circuits. And there is no comparing it at all to computers other than the fact that both are electronic devices and use similar components. Still, it is complicated enough in its own fashion, and that is why the process of "divide and conquer" is now applied. Turn on the radio. Does any sound at all come out of the speaker? If it does, then you immediately know several things about the device. First, you can assume that you have some power available because no noise would emanate from the speaker if the battery were dead or the power supply totally inoperational. You also know that the speaker is functioning as well. You are already in the process of dividing up the circuit into parts in an effort to determine which part to investigate further.

If n ɔ noise comes from the speaker, the first step is to get the voltmeter and see if there are any electrons available to make the circuit function. If you have just put a new battery in the radio and think that *cannot* be the problem, you just might be wrong. A dry cell has a shelf life of about one year. It can go dead just sitting on a shelf waiting for a customer. Even if you have just put a "new" battery in the radio, use the voltmeter to *verify* the existence of a power supply. Connect it to the battery and measure the voltage. Let us say that it reads + 9 V. Now turn the radio on. What does the voltage measure now? If this power supply is defective, the battery voltage will sag. If it drops more than a volt, the battery, although "new," may be useless. There also may be a short circuit, or an extra heavy load current may be being drawn. Try another battery or another power source. If the second battery does not sag, the first battery was definitely defective and will have to be replaced. If the voltage sags with a different battery, it is quite possible that an abnormal current is being drawn by the device under test.

Never forget that the power supply must be made operational first. Nothing in the device under test will function unless the correct power is applied. Once this hurdle is passed, a further splitting up of the circuit is in order. Try to "cut" the radio in half. The volume control is about half way through the circuit. Turn the volume control half way open, and hold a metal object in your fingers. Touch the center terminal of the volume control with the metal object. A good safety rule is to place your other hand on your hip or in your rear pocket.

Your body is surrounded by 60-cycle power lines. It will pick up a small amount of this 60-cycle voltage. When you touch the volume control, you will inject a small voltage into the audio amplifier circuits of the radio. This will be amplified by the circuits and heard in the speaker as a 60-cycle buzz. If you hear a buzz, you know that the back half, or output half, of the radio is functioning and that the trouble is in the "other half." If you don't hear anything al all, the first step is to find out what is wrong in the back half. The front half can be operational or not; it is unimportant at this point because even if it is working, the back half of the radio must be made operational in order to hear the front half.

Once you have split the radio into halves and determined which half to start investigating, you have your foot in the door. Test the speaker next. If the rest of the radio is operational but the speaker is defective, the radio will appear to be worthless. Find a way to inject a signal into the speaker to test it.

The principle of "divide and conquer" means to split the circuit in half and determine in which half the problem lies. This half is then split into half and the trouble isolated to one of the quarters. The process is repeated again and again until the trouble is found. It is a fundamental principle of troubleshooting—a simple concept that *can* result in a simple repair job.

Substitution

Another fundamental principle is that of substitution: You substitute a *known good component* for a suspected defective one. The problem in practice always turns out to be finding the *known* good component. Such a component turns out to be one that is known to work in the circuit, which brings us back to the introductory chapters of this book. Once you have an operational circuit, you then have known good components that you can use to substitute for suspected defective components. This is the reason that test circuits were given for all our experiments.

The principle of substitution is a sound one and one that you will use over and over again. We have already taken advantage of it by substituting another battery in the radio in the previous discussion.

Documentation

Another "fundamental" of troubleshooting is the schematic diagram or circuit diagram. Any paperwork available for a device will greatly simplify troubleshooting it. A discussion of its theory of operation may be included with the circuit diagram. The volume of paperwork that accompanies our computers is so extensive that we have given it a name: *documentation*. The better the documentation, the easier the job of getting things up and running will be.

Troubleshooting without a circuit diagram is extremely difficult but not totally impossible. If you are trying to repair a piece of electronic equipment and don't have a circuit diagram available, you should be made aware of Sams' Photofacts. These are circuit diagrams and repair aids (documentation) available at electronic supply houses. Most commercially manufactured pieces of equipment have documentation available through this source.

If no circuit diagram is available, you can do it the hard way. You can trace out the circuit and create your own circuit diagram. This is very time consuming and difficult, but it is a definite possibility.

Eyeballing

One troubleshooting technique is to "look for the obvious." When you have verified that the proper power is getting to the components, eyeball the circuit and look for things like missed solder connections, reversed diodes and capacitors (those polarized components), and loose or poorly connected wires. In soldering densely packed circuit boards, the solder may accidentally connect two adjacent traces or terminals, thereby creating a connection. This is called a *solder bridge* and can be extremely difficult to find. A sharp-pointed instrument can be run between any suspicious areas. A few minutes of eyeballing can sometimes save many hours of troubleshooting time.

Another very common occurrence is the bent-under pin, which can also be very difficult to find. This is one of the reasons that the inputs and outputs of ICs are checked on the top side of the board right at the pin of the IC itself rather than on the circuit-board side. Removal of each IC one at a time and inspection of its bottom for bent-under pins is one solution. But use caution. You may bend under the same pin again or even additional pins, compounding the problem even further when you reinsert the IC in its socket.

IC insertion tools can help circumvent this problem, but the best prevention is to know what can happen and to use care to prevent it from happening. An IC insertion tool is a device that holds ICs firmly and squeezes the pins inward slightly so that the IC can be inserted into its socket with greater ease and fewer bent-under pins.

Continuity Testing

One use of an ohmmeter is for measuring continuity. Verifying that one circuit point does in fact connect to another circuit point is a common test in troubleshooting. Likewise, verifying that two traces running next to each other on a circuit board do not have continuity when the two were not intended to connect is another common test.

An ohmmeter can be used for continuity testing, but it is not a particularly convenient tool since the ohmmeter must be read with the eyes and the eyes should be concentrating on the circuit under test. Taking the eyes off the circuit to read the ohmmeter consumes time and can easily lead to errors.

The construction of an audio continuity tester (which we named the "Squawker") was detailed in the chapter on PC board construction. This is an extremely handy tool for its intended purpose, and we strongly urge you to build one. You'll wonder how you ever got by without it if you do.

Knowledge of Circuitry

To stress the need for understanding the purpose of a circuit and how it accomplishes this purpose might seem to be overkill in a chapter on troubleshooting. Yet many students dive into a circuit to get it fixed without taking any time at all to try and figure out what the circuit does and how it does it. Take a few minutes to determine the purpose of the circuit and then another few minutes to determine how the circuit accomplishes it before you start troubleshooting. Unless you can fix the circuit by eyeballing or by determining that it does not have power applied, you will eventually have to do this anyway. You can save a great deal of time by firming up your knowledge of how the circuit functions before starting troubleshooting.

Short Circuits

One of the problems that can occur in a piece of equipment is that the darned thing blows fuses because of a short in the power supply. As soon as you put in another fuse, it immediately blows. Even though you eyeball everything, you still cannot find the culprit. You could increase the size of the fuse until it does not blow, but in doing so you may ruin everything and have to construct an entire new power supply.

Figure 14-1 shows an anti-fuse blower that is ultra simple and can be used in just this type of situation. It is nothing more than an 100- to 500-W, 120-V lamp in series with the line cord of the power supply. As long as the

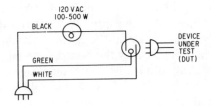

Fig. 14-1   The "anti-fuse blower circuit"

short is present that caused the fuse to blow originally, the lamp will be on at full brilliance. When the short is found and removed, the lamp will glow very dimly or not at all. The lamp should be used whenever a new power supply is constructed and initially tested. Any surprises awaiting you will turn on the lamp instead of making the newly constructed power supply go up in smoke. The lamp acts as a variable resistor in the circuit, and the output of the power supply will be a little lower than normal even if the lamp is not glowing. You can use the device to troubleshoot a shorted power supply when no other technique will suffice. Once the trouble has been found and corrected, the anti-fuse blower circuit can be bypassed and the power supply tested in its normal configuration.

Last Ditch Techniques

There are times when the problem simply cannot be found by conventional means. In the shorted power supply problem above, if you cannot track down the difficulty, you are justified in putting in a larger fuse or even jumpering the fuse and smoking the problem out.

Another difficult case is when two traces are bridged on a circuit board and you cannot find the bridge. The bridge may be burned open in a last ditch effort to correct the fault, using the circuit of Fig. 14-2. Here we charge up the filter capacitor with dc by momentarily contacting the power supply output contacts. We then remove all ICs and anything else removable from the circuit board and *very carefully* discharge the capacitor charge into the short. If this doesn't open the short, we add more capacitance, charge up,

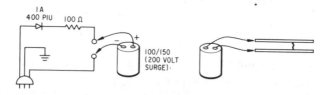

Fig. 14-2   The "blaster"

and try again. The circuit board will be useless if we cannot get the short out of the board. Sooner or later, traces will be blown or the short will vanish. Sometimes you can blast open the short without any damage to the circuit board at all. Other times, the circuit board will be ruined. Since the idea is to correct the fault and not destroy the board, the magnitude of the charge dumped into the short should be gradually increased until something gives. Remember that this is a last ditch effort that should be resorted to only after you have exhausted all other means of repair. The circuit is potentially lethal and must *never* be played with.

Micro-ohmmeters are said to be available for tracking down this kind of short, although the author has not seen one. A micro-ohmmeter would save much time and blown traces should one prove to be available.

Cutting Traces

The traces on a PC board are the wiring of the circuit. Consider the problem one faces when the ICs are soldered into the board and there is a short in the + 5-V line running to all of them. How in the world do you find out which one is shorted?

One technique is to cut traces. Trace the + 5-V line from the source of power through the circuit and all the ICs it feeds. Pick a point half way through the circuit, and run a knife blade through the + 5-V trace. Which half of the board is the shorted IC in? Verify that this trace is open with a continuity tester, and again try the power. Repeat the process until the defective IC is located and then replace it. Now patch the circuit board foil cuts with a short length of wire soldered across the cut. The circuit board will look "worked on" but will still be serviceable. Patience and skill are the keynotes here.

An alternative technique is the "lift and pull" method. The power pin is heated with a soldering iron and the power pin pulled up off the circuit board. It gets bent and mangled thoroughly and sometimes even breaks off, but the technique does not necessitate mangling the circuit board. Once the defective IC has been located and replaced, the task remains of opening up the holes under all the power pins that were lifted and of working the power lead back into the hole and reconnecting. Since this method requires a great deal of skill to keep from ruining ICs and circuit boards, the previous one is preferable.

Desoldering

This subject really does not belong in a chapter on troubleshooting; nor, for that matter, do some of the repair methods discussed above. But there are times when they are vital to troubleshooting techniques. Discussing this subject will let you see why most people recommend the use of sockets for ICs. Sockets can create contact problems, but they solve more problems than they create. Good quality sockets can prevent many hours of frustration in replacing soldered-in ICs.

When it becomes necessary to remove an IC or an IC socket, desoldering 14 or 16 pins (or even 100 pins on a mother board socket) simultaneously without damaging the circuit board can be a tremendous task. Special tools and techniques are called for.

Special soldering irons are available that have a hollow tip and a vacuum bulb attached to the iron. The tip is placed over the PC board connection, the solder is melted, and the vacuum device is used to remove it. The process is repeated until the pin is free of the hole. Use as little heat as possible to avoid damage to the board.

Another tool is a spring-operated plunger mechanism that is cocked and then fired with a trigger to create a vacuum. The tip of the device is hollow and made of Teflon® so that it will withstand the heat of an ordinary iron. An ordinary soldering iron is used to melt the solder, and then the tip of the desoldering device is placed very near the pin and the trigger pulled. The sudden rush of air into the device picks up the molten solder and removes it from the board. When the device is recocked, the solder residue picked up in the previous operation is ejected. All pins are freed and the IC removed for replacement.

Yet another tool for this purpose is the solder-wick. This is a piece of copper braid used to "wick-up" the molten solder. It does not get off as much solder as the vacuum devices, but it does help free the IC pins. Braid removed from a scrap piece of coaxial cable works about as well as the commercially available solder-wick.

In the desoldering process, the idea is to repair things and replace the defective components. Doing the least amount of damage possible to the circuit board and its traces is the goal. If an IC is known to be defective, it may be crushed with diagonal pliers and the pins removed one at a time. This is by far the easiest method on the PC board.

Logic Probes

A word about logic probes is in order. In Fig. 14-3, the 1000-ohm resistor on the two input 7404 sections pulls them low, thereby forcing both input section outputs high and turning off the upper LED (HI LED). The second inverter section then turns its high input upside down to turn on the LO LED. Since this LED is always on when power is applied, it can serve as a "pilot light" for the console. This constitutes a low impedance logic probe.

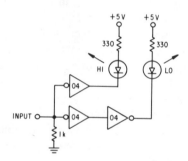

Fig. 14-3   A simple logic probe

In Fig. 14-4, we have a more elaborate circuit. Since the input circuit is a dual-emitter follower circuit, this is a high-impedance input circuit. Its input may be touched to voltages higher than + 5 V without being damaged. The balanced transistor pair forms a voltage splitter between + 5 V and ground. The junction of the two emitters rides at about 2.5 V. The 7404 input section considers this to be a high, and the two inverters in series hold the LO LED off. The lower NPN transistor has insufficient current into its base to allow its emitter to rise. The lower 7404 section (pin 9) is therefore pulled low by the 1000-ohm resistor on pin 9, and the HI LED is also off.

When a logic high is applied to the input, the upper transistor in the balanced pair turns on and the lower transistor turns off. Nothing happens to the LO LED circuit, and the LO LED remains off. The lower NPN transistor now has enough current into its base to be turned on. Its emitter rises, and pin 9 of the 7404 goes high, pin 8 goes low, and the HI LED turns on. When the input to the balanced pair goes low, the PNP transistor turns on and the NPN transistor turns off. The common emitter junction now goes low. Nothing happens in the HI LED circuit; the HI LED remains off.

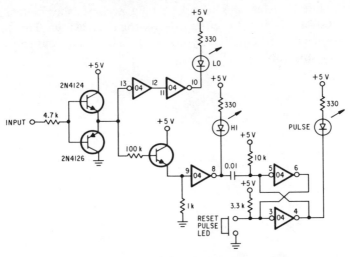

Fig. 14-4   A more sophisticated logic probe

However, the low on pin 13 of the 7404 places a low on pin 10, which turns on the LO LED.

The balanced pair may be any of the complementary pairs of NPN/PNP silicon transistors: the 4124/4126, the 3904/3906, and the like. If two transistors, one an NPN and the other a PNP, are connected as shown in Fig. 14-4 and the voltage at the common emitter junction measures about 2.5 V, the pair will work in this circuit. The third transistor may be any NPN silicon transistor as long as it has sufficient gain to switch in the circuit. In fact, students have built this circuit with totally unmarked transistors. Nothing is very critical in the circuit, and any values reasonably close will work.

This circuit also has a pulse catcher. The cross-coupled 7404 sections act as a latch. The pulse LED will be turned on and latched if a pulse is input to the circuit. The pulse LED will turn on if the input is taken high and will remain on until it is reset. It then serves as a pilot indicator and can also be used to capture a pulse that is far too narrow to illuminate the low or high LED.

The remaining single 7404 section can be used as a long-time constant "cheap shot" by employing a large capacitor for the coupling capacitor and stretching the pulses so that they can be seen with another LED added to the circuit. This would be a good design job for you. See if you can breadboard a circuit that will stretch the pulses wide enough to be seen with a LED.

Summary

Check the power supply first. Check the operating voltages right at the pins of the IC. Nothing will work right if the proper power is not applied.

Take the time to determine how the circuit accomplishes its task. Use the documentation. Eyeball the circuit. Look for the obvious. This precaution can sometimes save hours. Use your head. It is by far your most valuable troubleshooting tool.

Chapter 15

An Experimental Console

The modern solderless breadboard is an extremely powerful tool that can be used to test and design circuitry as well as to experiment with ICs. Figure 15-1 shows one method of incorporating the solderless breadboard into an experimental and educational framework that we shall call a *console*. Figure

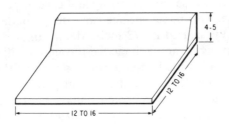

Fig. 15-1 Console sketch

15-2 shows the wooden framework used to construct the console, and Fig. 15-3 shows the completed console.

The wooden framework has been redesigned from earlier versions to make the carpentry as simple as possible. The console requires a sheet of ½-inch plywood, about 16 X 16 inches square. To this is added a sloping panel assembly about 4 to 5 inches high at the rear of the plywood base. The plywood shown has been covered with a plastic laminate to provide a smooth, hard work surface, and the Superstrip has been affixed to the work surface by using its self-adhesive backing. A suitable alternative to the plastic laminate would be particle board for the base. What is needed is a smooth, hard surface that will accept the Superstrip. Hardboard such as Masonite® will also prove to be an acceptable surface.

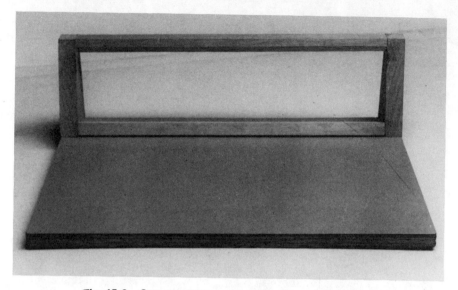

Fig. 15-2   One possible construction method for a console

The sloping panel at the rear is also covered with the plastic laminate. What is needed here is a thin, insulating board upon which to mount the support circuits that increase the versatility and ease of performing experiments on the console. This panel surface can also be made from Masonite if the maximum thickness used is 1/8 inch. Metal can also be used for the front panel, but every signal and voltage fed through it will then have to be insulated. All in all, the plastic laminate will prove to be the most satisfactory method of executing the base and the front panel.

The support circuits for the console include a built-in clock, a built-in logic probe, an SRFF (called the *start-stop control*) on the front panel, and a seven-segment LED read-out. Also recommended as a valuable accessory to the front panel is a card edge connector of the type used for plugging in electronic PC boards. Only power supply ground is connected to the card edge connector, but the connector can be used as a plug-in socket for additional circuits that may be needed later for special purposes. The connector thus helps keep the console from becoming obsolete.

Feed-through connections may be made in several different ways. One of the most rugged and least expensive methods of feeding signals and voltages through the panel is to use small nuts and bolts. If solder lugs are placed under the bolt head and under the nut and the combination is securely tightened, a most satisfactory feed-through may be created.

Another method of feed-through is to use IC sockets. These may be hacksawed apart and smoothed off to make satisfactory, high-density feed-through terminals for signals and voltages. They have an advantage over the nut/bolt/solder/lug combination in that connection wires may be simply

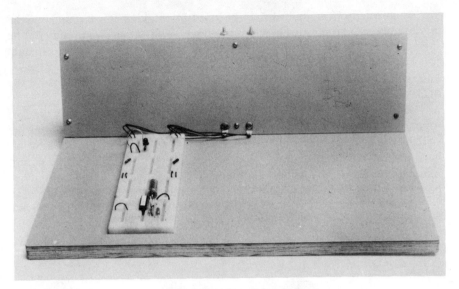

Fig. 15-3 The simple console

"plugged into" the IC connections to make easily removable connections between the front panel and the solderless breadboard.

Still another method of creating feed-through is to use Augat pin connectors. Pins can be pushed out of Augat wire-wrap boards and used for feed-throughs in the front panel. If the mounting hole for the Augat pin is quite snug, a simple wire loop soldered to the rear of the panel will "lock in" the Augat pin and hold it securely in place.

Fabricating the Support Circuits

There are different ways to fabricate the console support circuits. Consider Fig. 15-4, which shows the circuit for the start-stop control. The circuit needs power and ground and has two outputs: ground and the start-stop signal.

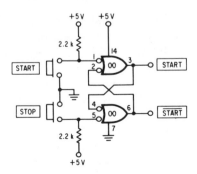

Fig. 15-4 The console start-stop control circuit

A printed circuit board may be used for making this circuit, but the circuit may also be simply haywired together. A haywired circuit may not be the most attractive electronic creation, but the end result is quite satisfactory. To haywire a circuit, wires are soldered directly to the pins of the IC and connections made to power, ground, and output. The IC hangs on the back of the panel, supported by the interconnecting wires.

Figure 15-5 shows the wiring diagram used to create this type of functional circuit. Let the wires that go to power, ground, output, and the switches be 6 to 8 inches long; they can always be shortened when the circuit is connected on the rear of the front panel. The functionality of the circuit may be verified on the solderless breadboard *before* it is placed behind the panel by plugging in the wires for power, ground, and output directly into

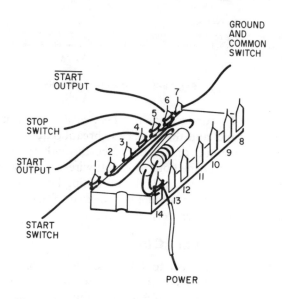

Fig. 15-5   Haywiring: one construction technique

the console breadboard. The switches are not needed for testing the haywired circuit since they simply make a connection to ground. The wires that go to the switches may be touched to the negative power rail on the breadboard to simulate the closure of the switches to ground. Once you have a proven circuit, it may be permanently connected behind the panel. It is a lot easier to troubleshoot "out front" than "out back."

Another alternative for getting circuits into operation without using PC boards is to drill the hole pattern in the front panel for a device and simply wire up the device on the rear of the panel. The soldered connections "lock" the device in place and hold it to the panel. The seven-segment readout may be mounted to the panel by this technique.

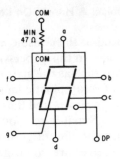

Fig. 15-6 The console seven-segment display

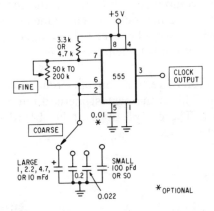

Fig. 15-7 The console clock circuit (a 7413 clock circuit may also be used)

The seven-segment read-out (see Fig. 15-6) is carefully pushed through a piece of paper. Be very careful because the pins of some read-outs are *very* easily bent. Doing so will establish the pin pattern to be drilled in the front panel. Drill the pattern with a small bit (#60 will be about right), insert the read-out into this set of holes, and make the connections on the rear of the front panel. Needless to say, the seven-segment read-out should be tested on the solderless breadboard *before* you make the effort of connecting it permanently. The current-limiting resistor needed with the read-out should be connected behind the front panel. It may be placed on the breadboard, but the first time you connect up the read-out directly between + 5 V and ground without it, you will blow the read-out. Since most of us are likely to forget to include the current-limiting resistor when using the read-out, it is prudent to solder it behind the front panel during the installation of the read-out. It can always be jumpered out of the circuit should this be necessary.

Depending on your choice of logic probes (Fig. 14-3 or Fig. 14-4), this console support circuit can be made operational and incorporated into the console. It may be haywired (Fig. 15-5) or hardwired (refer to the next paragraph) or made operational on a PC board. A console clock (see Fig. 15-7) may likewise be fabricated by the same techniques.

All circuits to be placed in the console may also be made operational by a process called *hardwiring*. In this process, the IC hole pattern is drilled in a piece of insulating board such as the glass epoxy board used to make PC board stock or in a piece of scrap plastic laminate.

The IC is inserted into the holes and the connections made on the back side of the insulating board. Small wire should be used; a suitable type is that used for wire-wrap. Instead of configuring the wiring of the circuit in a foil pattern, the wires themselves are connected to the circuit connections. Resistors and capacitors are likewise added by drilling a pair of holes in the board, inserting the component, and making the connections on the back side. The solder connections lock the components in place. This type of construction should be used whenever the circuit complexity is deemed too great for haywiring but not sufficiently complex to justify the fabrication of a PC board.

The final method of getting console circuits operational is to use PC board construction. This is the neatest method but also requires the most work on your part. The complexities of PC board construction will be covered in the next chapter.

Chapter 16

Printed Circuit Boards

The term *printed circuit* is a misnomer. In the earliest days of this type of circuit fabrication, conductors were actually printed onto an insulating substrate. Virtually all circuit boards used now are fabricated by etching away the undesired portions of the copper, leaving behind the conductors that interconnect the circuit. Since the term *printed circuit* (PC) has stuck with us, however, we will continue to use it, but we must realize that the process is a subtractive one (etching) rather than an additive one (printing).

The fabrication process begins with a piece of copper-clad board. A sheet of insulating material, such as Formica® or other plastic laminate or glass epoxy board, and a thin sheet of copper foil are bonded together. The thickness of the copper foil is measured in ounces (1-oz. copper foil is half as thick as 2-oz copper foil). You can manufacture your own copper-clad board by gluing copper foil to suitable insulating board. The result will probably turn out to be a far cry from the commercially produced copper clad, but the trick can be done. A suitable bonding agent is Goodyear Pliobond®.

No matter which method of PC board construction is selected, copper clad will be one of the necessary ingredients. A second necessity is an etchant, that is, a chemical solution that will slowly eat away the undesired portions of the copper clad. Ferric chloride is a common etchant. Since it is messy and stains very badly, it must be used cautiously. Ammonium persulfate is another etchant that is not nearly as messy as ferric chloride. A third chemical used for etching is cupric chloride. One of these three chemicals will have to be obtained if you are going to engage in PC board construction. All of them work more swiftly when heated; a lamp placed above, and directed toward, the solution is a simple source of heat.

A third requirement is a container to hold the etchant, sometimes called a *boat*. Metal should not be used since all three etchants attack metal. Glass or plastic boats are desirable. A very inexpensive, disposable boat can be fabricated from milk cartons. If a cardboard milk carton is used, slice it in two and retain the sealed half. Since excessive heat from the heat lamp will melt the wax used to make the milk carton waterproof, caution must be exercised in the application of heat. Plastic milk cartons may be used by simply slicing off the upper portion.

The fourth requirement for PC board construction is a *resist* A resist is any material that may be placed on the surface of the copper clad to resist the action of the etchant. The etchant eats away all the copper not protected by the resist, leaving behind the desired foil pattern.

The resist may a photographic resist. In the photographic process, a light-sensitive material is coated onto the copper clad. A photographic negative is made of the PC board pattern and used with a negative-acting photo-resist. A photographic positive is used with positive-acting photo-resist. The PC board is then exposed through this positive or negative, developed, and etched. The photographic method is ideal for reproducing many boards but is too expensive and time-consuming for producing single boards.

Materials are now available at local electronic supermarkets that may be applied directly to the copper clad, burnished (rubbed thoroughly) to bond them to the copper clad, and then etched. The applied materials serve as the resist. Contact Bishop Graphics at 5388 Sterling Center Drive, Westlake Village, CA 91359, for further details. These people also make all types of PC board construction aids, and you might want to ask them for additional information on these items as well. Bishop Graphics is also the source of Circuit Stik®, which is a series of thin copper-clad patterns already etched on a very thin glass epoxy base with self-adhering adhesive; it may be used to create hardwired circuit boards quickly and easily.

A resist pen may also be used to apply the resist pattern to the copper clad. This is a special felt-tip pen with "ink" that will resist the etchant. The resist pattern is drawn directly on the copper clad. Ordinary India ink will wash off in the etchant, but extra-dense India ink by Higgins has been used successfully. Nail polish is used by some PC board constructors. Any material, including lacquer and paint, that will resist the action of the etchant may be used for the resist.

The author's favorite is a material called *Bitumol* (pronounced "bitch-mole"). This is the same material that is sprayed on pavement by road maintenance crews prior to patching it. It is a water solution of bitumen, a coal product; when the water evaporates, tar is left behind. It is easy to apply with an old-fashioned ink pen if one or two drops of household detergent is added to the water solution to reduce the surface tension. Without the detergent, it is extremely difficult to get the Bitumol to flow off the pen point onto the copper clad. This material is very inexpensive but is also difficult to obtain. It is put up in 55-gallon barrels, a supply that would last you several centuries. Try bumming a small quantity from the boys on road maintenance crews.

Creating the Foil Pattern

The spacing for the ICs may be obtained from the pattern on the Superstrip. Place a piece of paper over the Superstrip and lightly shade the paper with the flat edge of a pencil point. Remove the paper, and with a small pin or needle, poke holes through the paper to establish the hole locations permanently. The shading may then be erased, with the hole pattern left behind.

Alternatively, place a piece of thin plastic such as a credit-card holder from a wallet or purse over the Superstrip and punch holes in the plastic to create a template. The template may be used over and over again if the holes are enlarged just enough to allow a pencil point to mark the hole pattern through it.

For all these tasks, you have been viewing the IC from the upper surface. When you view it from the bottom, as you would in creating the foil pattern for a circuit-board layout, the positions of the pins are reversed. Take an IC in your hand with the top toward you. Locate pin one, and put your finger on it. Keeping finger contact with pin 1, turn the IC upside down, and watch where pin 1 ends up. The first time you lay out a PC board pattern without paying attention to this reversal, you will instantly appreciate the overemphasis placed on it here. (If you should discover that you have inadvertently completed a PC board with the pins "read from the top," you can still salvage your work. The pins of each IC will have to be bent over opposite to their normal positions. When the IC is then inserted in the foil pattern, the circuit will function. It would be an extremely good idea to "flag" this error as a reminder—when you come across the board at a later date—of what happened and why all the pins seem to have the incorrect voltages on them.)

You should first lay out the PC board on paper so that changes and corrections may be easily made with an eraser. The circuit diagram should be placed on the upper half of the sheet. With the circuit diagram directly in front of you, you are less likely to make an error in laying out the board itself. Remember, that the process of laying out a board is equivalent to wiring up a circuit; the traces on the board serve literally as the connecting wires. Don't short-change yourself, therefore, by bypassing the step that transfers the circuit diagram from its source to your work sheet.

We will lay out the console clock circuit of Fig. 15-7 to illustrate the method used to create a PC board layout. Start with pin 1 and draw a small circuit around it with your pencil. This step is so vital that we have attached a small flag to pin 1 in Fig. 16-1. Now look at the circuit diagram to see what connections must be made to pin 1. Since pin 1 is shown grounded, we put in a pad to allow for this connection to ground, as shown in Fig. 16-2(a). We also add an arrow to the layout drawing to remind us that pin 1 will have to be connected to ground.

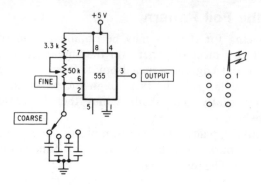

Fig. 16-1   Step 1: starting a PC board layout

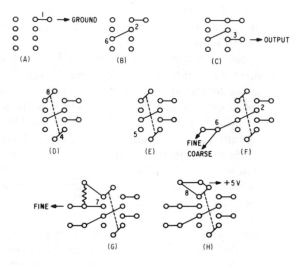

Fig. 16-2   Further steps in the PC board layout

Now we move on to pin 2. Again returning to the circuit diagram of Fig. 15-7, we see that pin 2 connects to pin 6. We place a pencil line on our layout drawing between pins 2 and 6, as shown in Fig. 16-2(b).

Next we tackle pin 3, shown going to the output pin in the circuit diagram. We put in a pad for this connection and another arrow indicating where it is to go, as shown in Fig. 16-2(c).

Pin 4 on the circuit diagram shows a direct connection to + 5 V. Pin 8 is also directly connected to + 5 V. We need to connect pin 4 to pin 8. We could go out and around pins 1, 2, and 3 to get to pin 8, or we could go around the other side to get to pin 8. We cannot go directly to pin 8 since the trace from pin 2 to pin 6 blocks that path. We will use the barricade from pin 2 to pin 6 to introduce a jumper [the broken line in Fig. 16-2(d)]. We will put in a pad next to pin 4 and another pad next to pin 8. The jumper can be a bare piece of wire. We will place this bare wire under the 555 IC with its ends bent down and soldered to the two pads provided for this purpose. The jumper is on one side of the insulating board; the foil trace from pin 2 to pin 6 that it crosses, on the opposite side. (Jumpers are often used in single-sided copper clad construction to cross over traces from one point to another.)

Pin 5 is next. Since the circuit diagram of Fig. 16-2(e) shows no connection to it, it poses no problem.

The next pin is pin 6, shown in the circuit diagram as being connected to pin 2. We have already made this connection. Pin 6 also goes to two other places. One connection is to the switch that connects in the timing capacitors. The second is to the fine clock speed control. Consequently, we must put in two pads next to pin 6. The arrows in Fig. 16-2(f) indicate that one connection will be made to the coarse frequency control and the second, to the fine frequency control.

Pin 7 is shown connected to two circuit elements also. One connection is to the fine frequency control, and the second, to a resistor whose other end goes to + 5 V. We thus place two pads next to pin 7, with an arrow to indicate the connection to the fine frequency control and a resistor symbol to indicate the connection to the resistor, as shown in Fig. 16-2(g).

Pin 8 connects to pin 4, to + 5 V, and to the resistor whose other end in connected to pin 7. The arrow in Fig. 16-2(h), indicates the wire that runs out to + 5 V.

The timing capacitors for the coarse frequency control must be mounted either on the coarse frequency switch or the PC board. Let's put them on the PC board. The circuit diagram shows one end of all these capacitors connected to ground and the other end to the terminals on the coarse selector switch. We will introduce another technique here. If all components are on hand during the layout stage, the distance between mounting pads can be determined by laying the components on the layout sheet and marking the location of the pads. If the components are not available at this time, we can do one of two things. We can find components later on that will fit the distance between the pads that we have already put in the layout, or we can add extra pads to accommodate varying sizes of components. The latter technique will be illustrated here.

Figure 16-3(a) shows pads for one end of the timing capacitors and also pads for the wires that connect the timing capacitors and the switch. Since we do not have the timing capacitors on hand, we place several mounting holes for the ground-end connections for each of them so that we can allow for varying capacitor lengths.

It takes time to etch away copper. The more copper that must be etched away, the faster the etchant will be exhausted. The more copper we can leave on the board, the less likely it is that the pads will separate from the board when we solder them on. Therefore, the idea is to leave as much

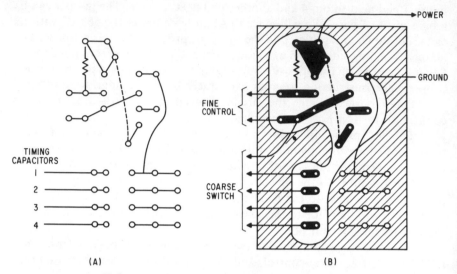

Fig. 16-3   Final steps in the PC board layout

copper on the board as we can and remove only that necessary to create an operational circuit.

Another useful technique is to place ground completely around a circuit, leaving the entire periphery of the board at ground potential and facilitating mounting with metal connectors such as bolts.

Figure 16-3(b) illustrates the principles just discussed. A line is run around the entire circuit to leave as much copper on the board as possible. The area shaded with cross-hatching is to be protected with resist. The pads near pin 8 that go to + 5 V are also interconnected.

Getting the Layout onto the Copper Clad

Now that the layout is completed, we must transfer it to the copper clad. First cut a piece of copper clad to the size of circuit board indicated by your layout. Use a piece of steel wool or very fine sandpaper and polish the copper clad until it is bright and shiny. Do not allow your fingers to come in contact with the polished surface. The body oil on them will transfer fingerprints to the surface, and oil of any type will prevent the resist from protecting the copper.

Hold the cleaned and polished copper clad against the back of your layout. Now hold the combination up against a light source. This can be an overhead light or a window in the room. Work the copper clad into position behind the layout, the copper surface against the paper. Now, without allowing either to move, return both to the work desk. With a sharp-pointed instrument (scribe, probe, needle, straight pin), transfer the center of each hole to be drilled (each pad location) to the copper clad with a pin prick.

Easy does it! All that is needed is the location, not a center punch! When all pads are transferred, remove the paper. Hold the copper clad in such a position that light will reflect from the surface and allow you to see every pin prick. With your pencil draw a small circle around each so that you will be able to locate them easily when the copper clad is returned to the work surface.

Now recreate the layout from the sketch using a pencil. (If you make an error, pencil lines are easily erased.Fingerprints on the copper clad will also give way to an eraser.) Once you have transferred the pencil layout to the copper clad and double-checked for errors, the resist may be applied.

Make the pads large enough for them to almost touch their neighbors. If they actually do touch, wait until the resist has dried; then with a sharp pointed instrument, scribe a line between the shorted pads. A common error is to make the pads too small. Remember that we are going to drill a hole through each pad, and we want some copper left after we drill it. If the pad is too small, no copper will be left. It is better to have the pad too large than too small.

Etching

Ferric chloride stains permanently, and caution must be used in handling it. The copper clad may be floated face down on the surface of the etchant if it is gently lowered onto it. When it is etched in this fashion, no agitation is needed, but you must be careful not to trap air bubbles under the board. Trapped air bubbles will prevent the etchant from reaching the copper and leave unwanted "pads" after etching. If the copper clad should sink to the bottom of the etchant, fish it out, turn it face up, and agitate the etchant. The copper clad will float as long as surface tension can be maintained. Once the board is wet, however, surface tension can no longer be maintained, and the board will not float. The board may be lifted for checking with a scrap piece of hook-up wire. If you use your fingers, immediately wash the etchant off them under running water. Do not use any of your tools, for the etchant will ruin them.

When the board has been completely etched (after about 20 minutes at room temperature using fresh ferric chloride), remove it from the etchant and flush it off with running water. If careful inspection reveals that any additional etching is needed, return the board to the etchant. If the board is properly etched, remove the resist. Steel wool will do the trick. (Bitumol may be removed with any petroleum solvent.) The board should now be polished so that it is again bright and shiny.

Drilling the Board

The etched board must next be drilled. If a great many holes must be drilled in epoxy board, then a PC board carbide drill should be obtained. Epoxy is a good abrasive and will quickly dull high-speed drills. A #60 drill bit is about the correct size for PC board work. Avoid too large a drill bit since larger holes make difficult soldering. Since small drill bits, especially carbide drill bits, break very easily, use caution. A drill press would be nice, but, with care, any drill motor can be used. Use a fairly high rpm for drilling.

If the board becomes contaminated with fingerprints during the drilling process, it must be cleaned and repolished.

Stuffing The Board

The process of inserting components into the board is called "stuffing" the board. Jumper wires *always* go in first since they often lie under other components. The flattest parts go in next to keep the board as flat as possible for soldering. The bulkier parts follow the flat ones. Each component is soldered in place individually, although on large boards all resistors may be soldered in place at one time. The wires that interconnect the PC board to the rest of the circuit go in last. Leave these 6 to 8 inches long. The board can be placed in the breadboard with these wires and tested before it is placed permanently behind the panel.

If you don't have a rotary switch for the coarse frequency control, simply bring out the two connections between pin 6 and ground, and run connections from these two points to the console breadboard. The timing capacitor can then be changed on the breadboard. The rotary switch is a convenient way to change the clock timing, but it is by no means mandatory.

The process described in this chapter takes a lot longer to explain than to execute. When only one-of-a-kind circuit boards are needed, it is the fastest way to get the job done and is entirely satisfactory for the simpler circuits. It may also be used for complex circuits after you have built up your skills with simpler ones. High-density circuits or very complex circuits are probably better left to photographic production techniques.

Some Hands-On Tools

Figure 16-4 shows an extremely handy tool that you can make by recycling a defunct ball point pen. The core of the pen is removed and discarded. A metal rod, such as a piece of metal coat hanger wire, is inserted

Fig. 16-4 A "hands-on" tool

into the plastic holder and glued in place with epoxy glue. One end is sharpened to a point, and the other end is rounded.

This device can be used as a scribe for working on PC boards, as a soldering aid (very handy for opening up the holes on PC boards), and as a

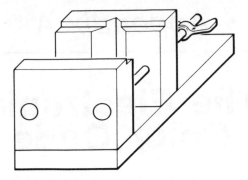

Fig. 16-5   Electronic work bench vise

probe attached to voltmeters, squawkers, and so on. It would be a good idea to make several because after you use the first one, you will want more.

Figure 16-5 shows an electronic bench vise. The end wooden block is firmly anchored in place. The second block is movable. The base is another wooden block. The vise action is provided by a pair of carriage bolts and a pair of wing nuts. The longer the carriage bolts, the better. When saw cuts (very shallow) are made in the two mating surfaces of the vertical blocks and a "V" groove is cut into one of the blocks (here shown in the movable block), the versatility of the vise is improved. PC boards may be held flat with the shallow saw cuts, and the "V" groove helps hold cylindrical objects (such as plugs) firmly for soldering. One of the author's friends said after he made his, "This has got to be the handiest device on my workbench!"

Appendix A

The Electronic Color Code

The value of resistors, given in ohms, is designated by the use of color bands. The value of capacitors, given in picofarads, is sometimes displayed with colored dots or colored bands. The color code is the same for all electronic components. The colors used are the colors of the white light spectrum (the colors of the rainbow) with black and brown added at the beginning and grey and white appended at the end. Thus,

Black	=	0
Brown	=	1
Red	=	2
Orange	=	3
Yellow	=	4
Green	=	5
Blue	=	6
Violet	=	7
Grey	=	8
White	=	9

Three bands are used to designate the value of a resistance in ohms (see Fig. A1). The band closest to one end of the resistor serves as the first digit. The second color band serves as the second digit. The third color band does not represent a digit, but rather a multiplier. It indicates the number of

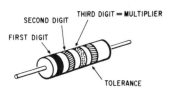

SECOND DIGIT

THIRD DIGIT = MULTIPLIER

FIRST DIGIT

TOLERANCE

Fig. A1   Coding for resistors

zeroes to be added after the first two digits. (A 101-ohm resistor could thus not be determined by color banding.) If gold or silver is used as the third multiplier band, they have a different meaning. Gold means that the first two digits are to be divided by 10 (multiplied by 0.1). Silver means that they are to be divided by 100 (multiplied by 0.01).

Resistors may contain a fourth color band. A fourth color band that is silver means that the value of the resistance given by the other three color bands has a 10 percent tolerance. A gold fourth color band means that the value has a tolerance of 5 percent. Likewise, the tolerance is 2 percent if the band is red, 3 percent if orange, and 4 percent if yellow.

A few resistors may have a fifth color band. The meaning of the fifth color band varies. It may refer to the temperature coefficient of the resistor or to the material from which the resistor is made. The meaning of this band is not significant for us.

Tubular capacitors marked like resistors are interpreted in the same fashion, the value being given in picofarads. Capacitors marked with colored dots are subject to different interpretations because several codes are used. In one standard code, the first dot is either black or white to identify a silver mica capacitor. The next dot serves as the first digit; the next after that, the second digit; and the last, the multiplier. The sequence is read clockwise in the direction of the arrow on the capacitor, starting with the black or white dot (see Fig. A2). Any other dots that may appear have to do with working voltages and temperature coefficients.

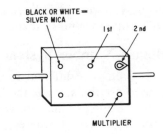

Fig. A2 Coding for capacitors

Appendix B

The Computer Number Systems

Most people have grown up with a number system that is based on ten different symbols—0, 1, 2, 3, 4, 5, 6, 7, 8, and 9—to express counting values from zero to infinity. It is called the *decimal system.*

Since digital circuits can have only two states, "on" and "off," electronic devices that count cannot make use of the ten symbols of the decimal system. They require only two different symbols, one to represent the "on" status, and another the "off."

In all counting systems, the position of a particular symbol is used to convey additional information about its value. In elementary school, this was called the *place value,* and you were introduced to the unit's column, the ten's column, the hundred's column, and so forth.

Binary Number System

A number system based on two different symbols is called a *binary system.* The two symbols used are the same as the first two symbols of the decimal system, 0 and 1. The meaning of the two symbols remains the same. A 0 means that we have nothing to count, whereas a 1 means that we have one item to count. A 0 for an electronic circuit is also interpreted to mean an "off" circuit condition and a 1 to mean an "on" circuit condition. For positive logic in working with digital circuits, the 0 is also called a low and the 1 is called a high.

In counting in the decimal system, we can count up to nine before we use all ten digits of the decimal system. Once we have used the ten different symbols to count to nine, we must use the same symbols again to express any greater value. We do this by using the 1 and the 0 again but place the two symbols in particular positions and when we position the two symbols in these particular positions, they take on a different meaning. A 1 placed to the left of a zero, as in 10, uses two of the decimal system symbols to convey the concept of "ten." Note that the sequence of writing the two symbols is important: 01 and 10 do not convey the same meaning although the same two symbols are used in each case. We call this concept *positional notation,* and we can now refer to the unit's column and the ten's column.

By using positional notation in which the symbol in the ten's column is multiplied by a power of ten, we can count to 99 in the decimal system before exhausting all the different combinations afforded us by the use of two different symbols. To express the next sequential count, we must utilize three symbols, as in 100, and we now have another power or ten column, the 100's column. By repeating this basic process, introducing an additional column based on a power of ten, we can express any number using just the ten different symbols of the decimal system.

The same procedure can be applied to the binary number system with its two symbols. The basic difference now will be in the values of the columns, which will be based on a power of two instead of a power of ten. Since we will be limited to just two different symbols, the "shifting of gears" in adding columns will occur much more often.

To illustrate, let us now count in the binary number system. Remember, we have only two symbols, 0 and 1, to use in expressing the values to be counted. If we have nothing, we represent this with a 0. If we have one item, we represent it with a 1.

We have now used the two symbols available to us to count with. To express the next count, we will have to do the same thing we did in the decimal system when we ran out of symbols. Namely, we will have to "shift gears." We must use the two symbols available to us and juxtapose them as 10. We have added a two's column, just as we added a ten's column in the decimal system. The symbol 10 now does not represent the decimal number ten but rather one item in the two's column and no item in the one's column. Therefore, 10 in binary is equivalent to 2 in the decimal system.

We have now counted two items. The next item counted adds a 1 in the unit's column to produce 11, which signifies one 2 and one 1, or 3. But we have run out of symbols already (again) and have to "shift gears" once more. To count the fourth item, we will have to add another column. We write 100 for this fourth item. This signifies one 4, no 2's and no 1's. The third column in the binary number system thus represents the 2^2 column, or the four's column. Note that each column added in binary is based on a power of 2 just as in the decimal system it is based on a power of 10.

We can now count to decimal 5 with 101, which is interpreted to mean one 4 and one 1. The next binary number, 110, means one 4 and one 2, or 6. The next is 111, which means one 4, one 2, and one 1. Again we have run out of symbols.

We now must add an eight's column (2 raised to the third power). We write this as 1000, meaning one 8, no 4's, no 2's, and no 1's. An 8 and a 1 make a 9, which we write as 1001. An 8 and a 2 make a 10, which we write as 1010. An 8, a 2, and a 1 make 11, or 1011 in binary. An 8 and a 4 make 12, or 1100. An 8, a 4, and a 1 make 13, or 1101. An 8, a 4, and a 2 make 14, or 1110. An 8, 4, 2, and a 1 make 15, or 1111. Again we have used all our symbols and must add another column in order to continue.

Binary 10000 represents decimal 16, or 2 to the fourth power. We can now continue counting until we run out of symbols again and have to add another column. The next power of 2 is 2 to the fifth, or 32, so that binary 100000 represents decimal 32, and so on. In this fashion counting values from zero to infinity are represented in the binary system.

Hexadecimal and Octal Number Systems

Note that the rather large groups of 1's and 0's are beginning to get a bit awkward to handle. A computer usually handles eight binary digits at a time, but a number like 10110110 is difficult for human beings to manipulate. One possible solution is to group the symbols in groups of three or four. We do this in the decimal system by using commas to break up large numbers into groups of three. Thus, 10110110 can be broken down into 1011 0110 to make it easier to work with, or it can be broken down into 10 110 110. Note that both of these groupings still represent binary numbers.

If we use groups of three and limit the number of symbols to eight, we call this octal notation.

If we add six additional symbols to the decimal number symbols, we can create a base 16 number system, called the *hexadecimal system*. It uses the ten symbols of the decimal system—0, 1, 2, 3, 4, 5, 6, 7, 8, and 9—and the first six letters of the alphabet—A, B, C, D, E, and F—to represent the numbers 10, 11, 12, 13, 14, and 15, respectively. The binary number 10110110, for example, is first divided into two groups of four, or 1011 0110, and then converted to hexadecimal as B6.

The important thing for you to grasp is that this new number system is just one way to make the binary numbers easier to handle. The octal system likewise is just a means to this end. The binary number 11101010 can be written as 11 101 010 and then converted to octal as 352, or it can be divided into groups of four, 1110 1010, and written as EA. Thus, 11101010, 352, and EA are all different ways of writing the same binary number. The octal system has the advantage of not requiring any new symbols, merely using the eight symbols of the decimal system, 0 through 7. These digits may be more conveniently displayed on a seven-segment readout. The hexadecimal notation uses fewer symbols to represent count values, breaks up binary numbers into two groups of four without anything left over, but it does require that the user learn a new number system and does not as easily adapt to the display of digits on a seven-segment readout.

Both number systems are commonly used in working with computers. If you had had as many years of practice working with either the octal or hexadecimal as you have had with the decimal system, you would find them both quite easy.

Table B1 compares the different systems with the more familiar decimal system. Use it as a reference as you work with the new systems, and it will not be long before your confusion clears.

Table B1   The Number Systems

Decimal	Binary	Octal	Hex
0	0	0	0
1	1	1	1
2	10	2	2
3	11	3	3
4	100	4	4
5	101	5	5
6	110	6	6
7	111	7	7
8	1000	10	8
9	1001	11	9
10	1010	12	A
11	1011	13	B
12	1100	14	C
13	1101	15	D
14	1110	16	E
15	1111	17	F
16	10000	20	10
17	10001	21	11
18	10010	22	12
19	10011	23	13
20	10100	24	14
21	10101	25	15
22	10110	26	16
23	10111	27	17
24	11000	30	18
25	11001	31	19
26	11010	32	1A
27	11011	33	1B
28	11100	34	1C
29	11101	35	1D
30	11110	36	1E
31	11111	37	1F
32	100000	40	20
33	100001	41	21
34	100010	42	22
35	100011	43	23
36	100100	44	24
37	100101	45	25
38	100110	46	26
39	100111	47	27
40	101000	50	28

41	101001	51	29
42	101010	52	2A
43	101011	53	2B
44	101100	54	2C
45	101101	55	2D
46	101110	56	2E
47	101111	57	2F

Appendix C

Listing of Parts

The following parts have been listed in chapter sequence. They are listed only once unless they are used to construct a support circuit; they are then assumed to be no longer available for experimental use and will be relisted for subsequent experiments. All the parts needed to complete the experiments in a given chapter are listed under that chapter. Unless otherwise stated, only one of each part listed is required.

Chapter 1

Superstrip or equivalent solderless breadboard
270-ohm resistor (180 to 470 ohms acceptable)
1.0-μF capacitor (0.47- to 6.8-μF acceptable)
Silicon diode (1N4000 series satisfactory)
Integrated circuit (basic TTL designation 7404)
Output coupling capacitor of 10 to 100μF
Permanent magnet speaker (size and voice coil Z insignificant[1])
Set of jumper cable components (a pair of alligator clips attached to
    each end of a piece of zip type lamp cord or speaker cable will be
    satisfactory; observe polarity in assembling jumper cable)

Chapter 2

Light Emitting Diode (LED)
Two silicon diodes (silicon power, silicon signal, or even germanium
    diodes acceptable)

Chapter 3

Two transistors, one npn and one pnp (power or signal; germanium or
    silicon)

[1] For school lab use, one or two PM speakers with the coupling capacitor affixed to the speaker cord will suffice for the entire class.

Open collector hex inverter (such as the 7405)
270-ohm resistor
2.0-μF capacitor (the value should differ from that of the capacitor in
     Chap. 1)
Additional coupling capacitor (10 to 100μF)

Chapter 4

7400 IC
Two SPST PB switches or an SPDT PB or toggle switch (switches are
     optional; jumper wires may be used to simulate their action)
100-μF capacitor (possibly the coupling capacitor from Chap.1)
7402 IC
Potentiometer (control or screwdriver adjust; standard or miniature;
     5000 ohms recommended)
LED
7408 IC (optional)
7432 IC (optional)

Chapter 5

7473 IC
Two current-limiting resistors (100 to 470 ohms)
Two LEDs
7410 IC
7474 IC

Chapter 6

7420 IC
7413 IC
1000-ohm potentiometer or rheostat
Quartz crystal (optional)
Two 470-ohm resistors (optional)
555 IC
4.7-k resistor
50- to 250-k potentiometer (control)
0.47-μF capacitor (the 1.0-μF capacitor from Chap. 1 may be sub-
     stituted)
74121, 74122, or 74123 IC
47-k resistor
4.7- μF capacitor

0.1-μF capacitor
470-ohm resistor

Following are components utilized in Chap. 6 and thereafter not available for experimental purposes:

Squawker:

555 IC
Two 4.7-k resistors
0.47-μF capacitor
10-μF capacitor
9-V transistor radio battery
Pair of alligator clips
Miniature speaker

555 variable clock:

4.7-k resistor
Control (50 to 100 k)
Four-position rotary switch
2.2-μF capacitor
0.1-μF capacitor
0.022-μF capacitor
100-pF capacitor
Two control knobs
555 IC

7413 variable clock:

7413 IC
1-k control
Four-position rotary switch
100-μF capacitor
0.1-μF capacitor
0.001-μF capacitor
100-pF capacitor
Two control knobs

1-MHz computer clock:

74123 IC
50-k miniature control
150-pF capacitor
50-pF capacitor
20-k resistor
1-k resistor

Chapter 7

7473 IC
7490 IC
7492 IC
7493 IC
Two 74161 ICs
74162 IC
7405 (any OC hex inverter may be used)
Two LEDs
Two current-limiting resistors (100 to 470 ohms)
74150 IC
1-k resistor
74190 or 74191 IC
74192 or 74193 IC
74196 or 74197 IC

Interval timer:

Four 74190 ICs
SPST PB switch
Two SPDT miniature toggle switches
7413 IC (only half circuit used; a partially good 7413 may be considered acceptable)
Three seven-segment decoders (type will depend on the seven-segment read-outs selected)
Three seven-segment read-outs (the FND 70 is about the least expensive available)
Current-limiting resistor (about 330 ohms)
LED (may be available in the seven-segment readout)
3.3-k resistor

Chapter 8

Two 74175 ICs
7474 IC
Two LEDs
Two current-limiting resistors (100 to 470 ohms)

Chapter 9

7488 (8223) IC
Two 7489 (8225) ICs
1-amp slo-blo fuse and fuse holder

Momentary-contact DPDT switch (PB or toggle; jumper may be substituted)

5000-μF/15-V capacitor (only one for an entire class)

Seven-segment read-out (the FND 70 is satisfactory)

Current-limiting resistor (47 to 470 ohms; try 150)

21L02 or equivalent RAM (a 1K RAM memory for a computer requires eight 21L02s or their equivalent)

Chapter 10

7442 IC

74154 IC

74139 IC

7446, 7447, 7448, or 7449 IC

Miniature LED assembly:

16-pin IC socket

Eight miniature LEDs

Eight 150-ohm resistors

Hot glue from a hot glue gun (glue gun is convenient but not essential; glue may be melted into position from the glue stick itself with a soldering iron tip, which must then be wiped clean)

Roulette type display:

Two 75154 decoders

Thirty-two LEDs

Thirty-two current-limiting resistors (100 to 470 ohms)

Two binary counters (7493, 74161, etc.)

74LS124 (plus timing components for same)

Chapter 11

74157 IC

74125 IC

74367 or its equivalent IC

74150 IC

0.01-μF capacitor

74151 IC

Hexadecimal keyboard encoder of Fig. 11-7:

Keypad salvaged from a defunct calculator (keys not arranged in a matrix)

Nonmultiplexed display of Fig. 11-8:

Six 7488 (8223) ICs (burned with appropriate decode pattern)
Six seven-segment read-outs
Forty-two pull-up/current-limiting resistors

Multiplexed display of Fig. 11-10:

Six 7488 (8223) ICs (burned with appropriate decode pattern)
7404 IC
270-ohm resistor
0.01-μF capacitor
Binary counter (7493, 74161, etc.)
7442 (or equivalent IC) used as 1:6 decoder
Seven pull-up/current-limiting resistors
Seven-segment display (salvaged from defunct calculator)

Multiplexed display of Fig. 11-11:

Four 8T97s (or their equivalent)
7488 (8223) IC (burned with common cathode decode pattern)
Two 7404 ICs
270-ohm resistor
0.01-μF capacitor
7442 (or equivalent IC) used as a 1:6 decoder
Binary counter
Seven pull-up resistors
Fairchild FNA45 nine-digit, seven-segment display

Circuit of Fig. 11-14:

Two additional 74161 counter/storage registers

Chapter 12

No additional components required.

Chapter 13

Power transformer (120-VAC primary to 12-VAC secondary at 1A)
Filter capacitor (the larger the better; 1000 μF at 15 working volts is about the minimum)
Four silicon power diodes (1N4000 series or equivalent)
Line cord

SPST toggle switch
½- to 1-A fuse and fuse holder
+ 5-V voltage regulator (7805, LM 340-5, LM 309K)
Zener diode (3 to 6 V or so)
Six silicon signal diodes (power diodes acceptable)
Silicon power transistor (almost any npn power transistor in any package acceptable; the 2N3055, a 15-A transistor, is quite reasonable in price)
7812 IC
7905 IC
7912 IC
120-VAC relay with DPST contacts
SPST PB switch, normally open contacts
SPST PB switch, normally closed contacts
200-ohm, 2-W control, screwdriver adjust
Two 10-A, slo-blo fuses and fuse holders
1000-μF/15-V capacitor

Optional components for special power supply circuits:

Chapter 14

Logic probe of Fig. 14-4:

Complementary pair of npn and pnp silicon transistors (the 2N4124/26 and the 2N3904/06 are a suitable pair, among others)
NPN silicon transistor (another 2N4124 or the equivalent)
7404 IC
Three LEDs (all red or different colors)
Three current-limiting resistors (about 330 ohms)
100-k resistor
4.7-k resistor
1-k resistor
10-k resistor
3.3-k resistor
0.01-μF capacitor
SPST PB switch

Circuit of Fig. 14-1:

Three-conductor line cord
120-VAC light bulb socket
120-VAC incandescent lamp (100 to 500 W)
120-VAC receptacle (outlet)

"Blaster" of Fig. 14-2:

400-PIV silicon power diode
100-ohm resistor
Three-conductor line cord
1000-μF (or more) capacitor at 150 working volts (or more) with a
    200-V surge voltage rating
Pair of jumpers permanently attached to capacitor

Chapter 15

Start-stop control:

7400 IC (only half of chip used; half-good chip acceptable)
Two 2.2-k resistors
Two SPST PB switches

Console clock for 555 circuit:

555 IC
50- to 250-k control with shaft
Four-position rotary switch
3.3- to 4.7-k resistor
0.01-μF capacitor (optional)
Four capacitors (10 μF at low end; 100 pF at high end)
Two knobs for the controls
Piece of copper clad (approximately 2 x 2 in.; type of board not
    important)

Console clock for 7413 circuit:

Control (1k with shaft)
7413 IC
Four-position rotary switch
Four capacitors (100 μF at low end; 100 pF at high end)
Two control knobs
Piece of copper clad (approximately 2 x 2 in.; type of board not
    important)

Console seven-segment display

Seven-segment display
One to three current-limiting resistors

Chapter 16

A supply of PC board stock
A supply of etchant
A material that can be used for resist
A suitable container to etch in
A supply of running water
Something to clean and polish PC stock with (very fine sandpaper or
    steel wool suggested)
Something to apply the resist with
A supply of graph paper ruled on 0.1-in. centers
A sharp instrument to use as a probe or scribe
An ordinary 2H pencil
An eraser

Appendix D

The Voltmeter

The voltmeter described in this appendix is the volt-ohm-milliammeter usually abbreviated as VOM. It is also called a *multimeter*, since it combines several measurement functions in one instrument. An attempt will be made to make the discussion applicable to all instruments of this type.

The basic instrument consists of a sensitive meter movement to which a needle-thin piece of aluminum is attached. When a very small current flows through the meter mechanism, the torque developed causes the needle or pointer to rotate. The greater the current, the more torque and the greater the displacement of the pointer from its resting, or zero, position.

The Meter Controls

Examine the front panel of your instrument. It may contain several switches or a single switch as its control. Some instruments do not use switches but separate jacks in which to insert the meter leads. The VOM has two leads, usually red and black, connected to the front panel. The red lead is the positive lead and the black lead is the negative lead.

If your instrument uses jacks rather than control switches, the connection of the positive lead (red) controls the instrument.

If your instrument has more than one control switch, one of them will be the *function* switch and the other, the *range* switch. The function switch is used to control the instrument for measuring voltage, current, or resistance. The range switch is used to select the scale upon which the measurement will be made.

If your instrument has a single control switch, both the function and the range are controlled with this single control.

Safety Positions

Each measuring instrument has what is called *safety positions*. For all VOMs, these would be the highest range and ac. Your instrument may also have a function called *output*. When the controls are placed in the output position, a capacitor is placed in series with the positive lead, and the instrument operates on ac. This would be the appropriate safety position for such an instrument.

If your VOM does not have an ac function, then the appropriate safety positions are the highest range and dc. If dc+ and dc- are both available, use the highest range and dc+ for the safety positions. Form the habit of leaving your instrument in the safety positions at all times. Never leave it in the ohms function when you are finished using it.

Measuring DC

Place the instrument in the position of use for reading the measurement. Your eye should be at right angles to the meter scale to make this measurement. Some meters have a mirrored scale. Position your eye so that you see only one pointer image and cannot see the pointer reflected by the mirror. Connect the two meter probes together. The instrument should read 0 V. If the needle does not indicate 0 V, it needs adjustment. Some instruments have a screwdriver adjustment on their face at the point where the pointer pivots. This is the mechanical zero adjustment, and the needle should be zeroed with it. If your instrument does not have this mechanical zero adjustment, then you must position the instrument so that it reads zero to your eye.

Range Selection

If you know the approximate voltage to be measured, set the range switch in the appropriate range for this voltage. If you have no idea of the voltage to be encountered, then set the range switch in its *highest* range. The common lead (black or negative) is connected to the more negative point in the circuit. This will usually be the chassis or ground of the equipment under test. Pick up the positive probe and hold it by the insulating handle, with your other hand on your hip or in your rear pocket. Place the probe tip on the circuit point where you desire to measure the voltage, and read the voltage on the VOM scale.

Selecting the Correct Range

Let us assume that you had no idea of the voltage to be encountered and had initially set the voltmeter in its highest range. The needle may or may not move in your first measurement. Set the probe down and switch the VOM to its next lower range. Again make the measurement. The needle should swing upscale. If the needle is still very close to zero, put the probe down and change to the next lower range. You will eventually get the needle into the center of the range of the instrument and should make the reading at this point. If the needle goes off scale at the right, you went too far and need to select the next higher range.

If the needle attempts to go below zero, the current through the meter mechanism is going in the wrong direction. The negative lead of the instrument is at a point more positive than the point where you are placing the probe. Reverse the meter leads. Ground the probe and measure the voltage with the negative clip or lead. If your instrument has dc+ and dc— switch positions, you need only change this switch to dc— and measure with the red lead. This is the purpose of this switch—to reverse the meter leads for you.

Reading the Scales

The newcomer usually has difficulty reading the scales on the meter. The multimeter has several scales. One scale is used to read resistance in ohms. The ohms scale may have the zero at the left or at the right.

One or more scales will be used to measure dc. The same scales may be used to measure ac, or the multimeter may have separate scales for ac. The scales for dc may also be used to measure milliamperes (and even amperes). Study the scales on the instrument you are using, and see if you can determine which scale is used for each function.

Now examine Fig. D1, which depicts a generalized VOM. The uppermost scale on this meter is the ohms scale. Its use is controlled by the

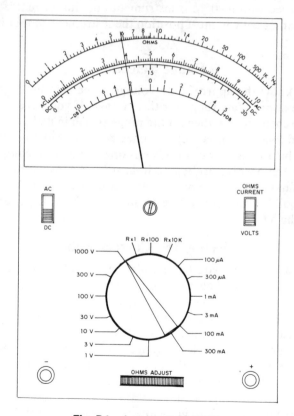

Fig. D1   A generalized VOM

large knob in the center of the meter and the slide switch below the meter and to the right. The ohms scale is read directly when the large knob is in the R X 1 position. If the large knob is in the R X 100 position, two zeroes would be added to the reading on the ohms scale. If the large knob is in the R X 10K position, four zeroes would be added to the reading on the ohms scale.

Voltages and current are read on the two middle scales marked ac/dc. If the large knob is in the 30-V position and the slide switches are set to volts and dc, the dc voltage would be read on the 30-V scale (the lower of the two center scales). If the large knob is set to the 10-V position, the measurement would be made on the upper scale of the two. Both of these would read directly. If the large knob is turned to either the 100-V or the 300-V position, the voltage read on the corresponding scale would need one zero appended to the reading. If the large knob is in the position shown in the diagram, the voltage would be measured on the 10-V scale at the top of the two, and two zeroes would be added to the reading. The needle rests on 3.9 so that the voltage shown in the diagram would be 390 V. If the large knob is on the 3-V or the 1-V position, the reading on the 30-V or the 10-V scale would have to be divided by 10. On the 1-V range, the meter indicates 0.39 V. AC voltages would be measured by placing the left slide switch in the ac position.

Voltages are measured in parallel, whereas currents must be measured in series. To measure the current flowing in a circuit, the circuit must be opened somewhere in the current loop that you desire to measure and the meter inserted in series with the electron flow of the circuit. Polarity of the circuit and the meter must be observed. Electrons must flow into the meter on the negative multimeter lead and back into the circuit on the positive meter lead. Connections of the multimeter into the circuit for measuring current must always be made with the line cord to the equipment under test unplugged. At the start of the current measuring test, the multimeter should always be set at the highest current range available (in this diagram, the 300-mA range). In making a current measurement, the left slide switch is set to the dc position and the right slide switch to the ohms/current position.

The decibel scale at the bottom of the drawing would be used for making power measurements. For its proper use, refer to the instruction manual that accompanied your multimeter.

Calibration

The newcomer to electronics usually tends to accept meter readings at face value. Most of the time, this attitude is acceptable, and especially with new equipment. To be able to "trust" what your instrument is telling you requires that it be periodically calibrated against a known standard or standards.

The voltmeter scales of an average VOM can be tested on dc by using a mercury cell battery. Each cell of the mercury cell battery has a voltage of 1.34 V and this voltage remains constant over quite long periods of time (five years or more). When the mercury cell becomes exhausted, it dies almost overnight; so you can therefore usually depend on its voltage to be either 1.34 or 0.00 V. An 8.4-V mercury cell battery would make a suitable calibration source for the instrument depicted in Fig. D1.

Calibration on ac is much more difficult than on dc. Since the ac line voltage available at a wall outlet can fluctuate widely, it should not be used as a calibration source. If your multimeter checks out okay on this dc calibration check, you can reasonably assume that it is still just as accurate as it was when you got it. There is no guarantee of this, of course, but it is a reasonable assumption.

Calibration of the ohmmeter function is relatively straightforward. You need to obtain precision (1-percent tolerance) resistors and check the readings against your instrument. Try to obtain resistance values that would produce about half-scale deflection of the needle. For our generalized instrument, these would be 10, 1000, and 100,000 ohms.

Calibration of the current functions is not easy. Unless you use the current function often, it really does not pay to bother. I think that you will find your instrument used about 90 percent of the time for voltage measurements, about 10 percent of the time for resistance measurements, and much less than 1 percent of the time for current measurements.

The simple circuit of Fig. D2 can be set up to form a current calibration circuit that will let you know if your instrument is in the "ball park." Obtain a precision (1-percent) resistor in the vicinity of 10.0 ohms. Use a

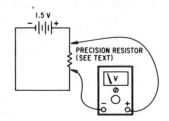

Fig. D2 Setting up for the current calibration check

standard flashlight cell for the power source. Check the calibration of your instrument on dc volts as outlined earlier before making the current calibration test. Connect up the circuit of Fig. D2, measure the voltage across the resistor (and not the voltage across the battery), and record this voltage. Do this as quickly as you can as the current flow through the resistor will heat the resistor and change its value. Now disconnect the circuit of Fig. D2. Place your meter in the appropriate current range. For the demonstration instrument shown in Fig. D1, this would be the highest range, or 300 mA. Switch the controls to the current position, and insert the meter into the circuit in series as shown in Fig. D3. Quickly measure the current indicated by the meter, disconnect the meter, and record this current.

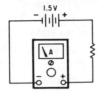

Fig. D3 Inserting the meter for the current check

Now Ohm's Law can be used to calculate the current that you should have measured. Ohm's Law states that the current in amperes equals the voltage divided by the resistance $(I = E/R)$. Let us assume that you have a 10.0-ohm resistor and the voltage drop across the resistor measures 1.48 V. Plugging these values into Ohm's Law would produce the following:

$$I = \frac{E}{R}$$
$$= \frac{1.48}{10.00}$$
$$= 0.148 \text{ A}$$

Dividing this by 1000 to convert amperes to milliamperes produces 148 mA. Compare this with the current that you measured. You should find that you measured a slightly smaller current because the meter itself has a small amount of internal resistance when used to measure current and this resistance is added to the 10.0 ohms placed in the circuit. Since this added resistance will be on the order of 0.01 ohm, for all practical purposes we can ignore it. Since we cannot calibrate our instrument but only verify its calibration, we can obtain only "ball park" figures with our simple equipment.

Meter Life and the Rules of the Road

The VOM is a very sensitive piece of equipment and does not survive long in a novice's hands unless he or she uses it carefully and observes adequate precautions. A VOM will last for a long time in the hands of an experienced user. To overcome the lack of experience, here are some "Rules of the Road."

1. Use the safety positions for your instrument. Form the habit of always returning its controls to these positions.

2. Always set all controls to their appropriate positions *before* picking up the test probes to make the measurement.

3. Use care to avoid dropping the instrument (concrete floors are very hazardous to its health).

4. Use one hand only when making measurements. Place the other on your hip or in your rear pocket. You might wonder how this will help the meter when it is a safety rule for your own health. If you could see a meter flung across a room when a user reacts to severe electrical shock, you would quickly appreciate the fact that this rule will help protect the meter as well as the operator.

5. *Never* use an ohmmeter on a circuit that has power applied. This ruins more instruments than any other error. Do not just turn the equipment off with the on-off switch, but disconnect the line cord from the equipment as well. It is also advisable to discharge the filter capacitors in the power supply as well. This cannot be done with power applied. With power disconnected, a length of wire can be used to short the positive and negative terminals of the filter capacitors together to remove any stored charge. Many ohmmeters get zapped because of this stored energy that remains even after the line cord has been disconnected.

Safety

Measuring high voltages with your equipment is the most dangerous. You should consider every voltage measurement as a potentially lethal measurement. It would be difficult to overemphasize this point. Use care and *think!* The life you save could be your own!

Index